SEM Micrograph of the
complex trabeculated
inner side of the apex
in rat right ventricle
from
Endocardial Endothelium:
Functional Morphology
by
Luc J. Andries
© R.G. Landes Co 1994, 1995

Received 31/5/03
sent by my sister
Doreen about 1mth ago

MEDICAL
INTELLIGENCE
UNIT

ADULT T CELL LEUKEMIA AND RELATED DISEASES

Takashi Uchiyama, M.D.

Kyoto University
Kyoto, Japan

Junji Yodoi, M.D.

Kyoto University
Kyoto, Japan

Springer-Verlag
New York Berlin Heidelberg London Paris
Tokyo Hong Kong Barcelona Budapest

R.G. LANDES COMPANY
AUSTIN

MEDICAL INTELLIGENCE UNIT
ADULT T CELL LEUKEMIA AND RELATED DISEASES

R.G. LANDES COMPANY
Austin, Texas, U.S.A.

U.S. and Canada Copyright © 1995 R.G. Landes Company
All rights reserved. Printed in the U.S.A.

Please address all inquiries to the Publisher:
R.G. Landes Company, 909 Pine Street, Georgetown, Texas, U.S.A. 78626
or
P.O. Box 4858, Austin, Texas, U.S.A. 78765
Phone: 512/ 863 7762; FAX: 512/ 863 0081

U.S. and Canada ISBN 1-57059-195-4

International Copyright © 1995 Springer-Verlag, Heidelberg, Germany
All rights reserved.

International ISBN 3-540-58895-7

While the authors, editors and publisher believe that drug selection and dosage and the specifications and usage of equipment and devices, as set forth in this book, are in accord with current recommendations and practice at the time of publication, they make no warranty, expressed or implied, with respect to material described in this book. In view of the ongoing research, equipment development, changes in governmental regulations and the rapid accumulation of information relating to the biomedical sciences, the reader is urged to carefully review and evaluate the information provided herein.

Publisher's Note

R.G. Landes Company publishes five book series: *Medical Intelligence Unit, Molecular Biology Intelligence Unit, Neuroscience Intelligence Unit, Tissue Engineering Intelligence Unit* and *Biotechnology Intelligence Unit*. The authors of our books are acknowledged leaders in their fields and the topics are unique. Almost without exception, no other similar books exist on these topics.

Our goal is to publish books in important and rapidly changing areas of medicine for sophisticated researchers and clinicians. To achieve this goal, we have accelerated our publishing program to conform to the fast pace in which information grows in biomedical science by publishing most of our books within 90 to 120 days of receipt of the manuscript. We also like to hear from our readers—their interests, comments and suggestions for future books.

Deborah Muir Molsberry
Publications Director
R.G. Landes Company

ACKNOWLEDGMENTS

We thank all of our former and present collaborators in writing this monograph. Particularly, we express our deep appreciation to Drs. Kiyoshi Takatsuki, Thomas A. Waldmann and Kimishige Ishizaka.

CONTENTS

Preface 1
    Junji Yodoi
    Adult T Cell Leukemia (ATL) 1

1. **History** 5
   Junji Yodoi
   Initiation of the ATL Study in Japan 5
   Independent Study of CTCL and ATL in the U.S. 6
   Mechanism of ATL Leukemogenesis Related to IL-2R Dysregulation ... 9

2. **Clinical Features of ATL and Other HTLV-I-Related Diseases** 17
   Takashi Uchiyama
   ATL 17
   Other HTLV-I-Related Diseases 29

3. **Human T Cell Leukemia Virus Type I (HTLV-I)** 39
   Junji Yodoi
   Retrovirus 39
   Human T Lymphotropic Virus Type I (HTLV-I) 42
   Human T Lymphotropic Virus Type II (HTLV-II) 43

4. **Biological Properties of ATL Cells** 45
   Takashi Uchiyama
   Cell Surface Phenotype and Function In Vitro 45
   Cytokines and Other Biologically Active Peptides
       Produced by ATL Cells 50

5. **IL-2/IL-2 Receptor System in ATL** 59
   Takashi Uchiyama
   Interleukin-2 59
   Interleukin-2 Receptor (IL-2R) 64
   IL-2/IL-2R Expression in ATL 69

6. **Proliferation of ATL Cells In Vivo** 81
   Takashi Uchiyama
   Engraftment of Fresh Leukemic Cells from ATL Patients
       in SCID Mice 87
   Cell Growth of HTLV-I-Infected Cell Lines in SCID Mice 91
   Radiometric Studies on the Proliferation
       of ATL-43 T Cells In Vivo 95

7. Redox Regulation and ADF 101
Junji Yodoi
ATL-Derived Factor(ADF)/Human Thioredoxin (hTRX) 101
Redox Dysregulation in ATL and Other Viral Infections 110
Redox Control of Signal and Gene Expression 115
Virus-Related Transformation 120
Serum ADF/TRX 125
Clinical Utility of Redox Control and ADF/TRX 126

Index 133

PREFACE

Junji Yodoi

ADULT T CELL LEUKEMIA (ATL)

THE HISTORICAL ROLE OF ATL AS THE FIRST DESCRIBED HUMAN RETROVIRAL DISEASE

We first encountered a group of patients with abnormal lymphoid cells characterized by convoluted nuclei and T cell properties in Kyoto University Hospital in the early 1970s. Today, this disease entity is called adult T cell leukemia (ATL).[1,2] *ATL is one of the central disorders related to infections caused by the retrovirus called human T lymphotropic virus (HTLV-I).*[3] *Because of its direct relationship to the study of ATL, the HTLV virus was also once called ATLV in Japan.*[4,5]

We will first review the basic and clinical aspects of ATL and related disorders, considering the rapid expansion in knowledge of human diseases related to retroviral infections since initiation of the ATL study.

This monograph provides an overview of the study of ATL as the first human retroviral disease, as well as the current knowledge on clinical and basic aspects of ATL and related diseases. We will also discuss the future direction of basic and clinical efforts to control these diseases.

Another retroviral disorder, acquired immune deficiency syndrome (AIDS) which is related to the human immunodeficiency virus (HIV), is briefly compared to the relationship between ATL and HTLV-I.[6]

THE IMPORTANCE OF ATL IN THE CURRENT STUDY OF RETROVIRAL DISEASE

There are several unique features of ATL and related disorders caused by HTLV-I. First, ATL was the first retroviral disease discovered in humans. As the first example of HTLV-I-related disorders, the discovery of ATL facilitated the discovery of other disorders related to this virus and the recognition and study of another retroviral disorder, AIDS. Second, ATL was one of the first direct examples of virus-related human leukemia in the history of modern cancer research to prove the viral etiology of malignant diseases. The third important feature is the multiple disease phenotype of disorders following HTLV-I infection.

The clinical course of ATL encompasses a spectrum which includes aggressive acute subtype ATL (acute ATL) to chronic phenotype ATL (chronic ATL). Clinicians have proposed intermediate groups of ATL such as subacute and smoldering types based on distinct clinical manifestations among subtypes, which may now be partially explained molecularly.

An important principle we have learned from the study of ATL is the fact that HTLV-I infection may lead not only to malignant disease such as ATL, but also to various non-malignant organ- or tissue-specific disorders. These non-malignant disorders, tentatively called HTLV-I-related diseases, seem to involve more or less immunological dysfunction or autoimmune states and are discussed later in this monograph.

IMMUNE MACHINERY DISTORTED RETROVIRAL INFECTION

There are a variety of immune dysfunctions in HTLV-I infection including ATL and non-malignant disorders. As a T cell tropic virus at the cellular level, HTLV-I infects helper T cells having CD4 antigen (TH). In ATL, leukemic cells are generally $CD4^+$. Unlike T cells from normal individuals, T cell lines are relatively easily to establish from peripheral blood lymphocytes (PBL). They are generally $CD4^+$ T cells and are infected with HTLV-I.

An interesting feature of ATL cells is their abnormal activation phenotype with overexpression of Tac antigen associated with α chain of interleukin-2 receptor (IL-2R). Another ATL cell characteristic is augmented production of a variety of cytokines and other inducible proteins including ATL-derived factor (ADF).

STRESS OF THE IMMUNE SYSTEM IN RETROVIRAL DISEASES

The concept of immune system stress is a crucial element in the study of how retroviral diseases affect T cells. In ATL, activated T cell phenotype with IL-2R α chain overexpression indicates continued activation of the infected T cells by either extracellular or intracellular hypothetical activators. The close relationship between continued activation and T cell stress is indicated by the fact that HTLV-I⁺ T cell lines constitutively produce ADF, an oxystress-inducible anti-stress protein with many biological activities.

Loss of TH cells, which is partly explained by the enhancement of cell death, is one of the cardinal features of AIDS. Involvement of the oxystress generated by cytokines and HIV has also been speculated. Chronic infection of T-tropic retroviruses such as HTLV-I and HIV, therefore appears to cause sustained stress on the host immune system.

HOST AND VIRAL GENES

ATL develops only in a limited proportion of individuals infected with HTLV-I, generally in aged individuals, which indicates the presence of a host defense mechanism against the leukemogenesis. The spectrum of diseases developing in HTLV-I-infected populations illustrates the importance of studying the host immune system against these infectious agents.

REFERENCES

1. Yodoi J, Uchiyama T. IL-2 receptor dysfunction and adult T-cell leukemia. Immunological Review 1986; 92:136-56.
2. Yodoi J, and Uchiyama T. Human T-cell leukemia virus type I (HTLV-I) associated diseases; virus, IL-2 receptor dysregulation and redox regulation. Immunology Today 1992; 13:13:405-11.
3. Poiesz BZ, Ruscetti FW, Gazdar AF et al. Detection and isolation of type C retrovirus particles from fresh and cultured lymphocytes of a patient with cutaneous T cell lymphoma. Proc Natl Acad Sci USA 1980; 77:7415-9.
4. Hinuma Y, Nagata K, Hanaoka M et al. Adult T-cell leukemia; antigen in an ATL cell line and detection of antibodies to the antigen in human sera. Proc Natl Acad Sci USA 1981; 78:6476-80.
5. Yoshida M, Miyoshi I, Hinuma Y. Isolation and characterization of retrovirus from cell lines of human T-cell leukemia and its implication in the disease. Proc Natl Acad Sci USA 1982; 79:2031-5.
6. Gallo RC, and Streicher HZ. AIDS Modern Concepts and Therapeutic Challenges. Broder S, ed. New York: Marcel Dekker Inc, 1987:1-21.

= CHAPTER 1 =

HISTORY

Junji Yodoi

Around 1970, the development of cellular immunology based on the battery of mice strains led to consensus about the T and B cell dichotomy of the immune system. Prior to introduction of monoclonal antibodies, polyclonal antibodies specifically reactive with T cells were used to analyze immune mechanisms. We attempted to obtain a specific anti-T cell antiserum for the study of human lymphoproliferative disorders.

INITIATION OF THE ATL STUDY IN JAPAN

DISCOVERY OF A PECULIAR T CELL LEUKEMIA IN JAPAN

In the early 1970s, the first case of ATL was reported in Kyoto, Japan. In a cooperative study with Dr. Masuda in the Virus Research Institute of Kyoto University, Yodoi and Takatsuki prepared a rabbit antisera against human T cells, by the successive absorption of rabbit anti-human thymocyte antiserum with various non-lymphoid cells. Using the relatively specific anti-T cell antibody and the well-known rosette formation with sheep red blood cell (SRBC) as a T cell marker, we identified a patient with chronic lymphocytic leukemia (CLL) of T cells.[1] The female patient was from Oki-No-Erabu-Jima, an isolated southern Japanese island.

Subsequent study of similar cases, in cooperation with other university hospitals in Kyushu and elsewhere, revealed that the majority of patients with leukemic cells of T cell phenotype originated from southern Japan, particularly Kyushu, Shikoku and related small islands (Fig. 1.1). Based on these findings, we proposed that

Adult T Cell Leukemia and Related Diseases, edited by Takashi Uchiyama and Junji Yodoi. © 1995 R.G. Landes Company.

adult onset T cell chronic leukemia was frequent in Japan, as compared to the prevalence of B cell CLL in the U.S.[2]

PATHOLOGY OF ATL

In the late 1970s, the study of ATL was well-established in the fields of pathology and hematology. This clustering of clinical cases in southern Japan attracted the interest of various investigators in Japan and other countries. The heterogeneity of their clinical status, morphology of abnormal T cells, and the infiltrating tendency of the skin and other tissues were recognized as unique features of the disease.

It is known that a typical morphological feature of ATL cells is a convoluted nucleus similar to other T cell malignancies such as mycosis fungoides and Sezary's syndrome. The mechanism of the infiltrating tendency and the origin of the abnormal cells in relation to the skin, however, remains to be clarified.

CELL LINES FROM ATL

Various trials to establish leukemic cell lines were unsuccessful until the mid-1970s, as fresh leukemic cells from ATL patients were difficult to maintain in in vitro culture.

Ishii et al[3] first established a hybridoma cell line between ATL cells and non lymphoid cells, although the phenotype analysis of the hybridoma was incomplete. Subsequently, Miyoshi et al reported MT-1 cells established by the co-cultivation of ATL leukemic cells and umbilical lymphoid cells. Despite the lack of T cell properties such as CD4 common to the leukemic T cells, MT-1 cells contributed to the initial work in the discovery of the associated virus in Japan.[4]

INDEPENDENT STUDY OF CTCL AND ATL IN THE U.S.

In the late 1970s, several lines of work at the U.S. National Institutes of Health (NIH) opened a new chapter in the study of ATL and related diseases and the virus associated with these diseases.

T cell clones had been established by several NIH research groups from a patient with cutaneous T cell leukemia (CTCL).[5] The disease diagnosis of the patient was later correctly changed to ATL. The causative virus was discovered in the laboratory of Gallo

Fig. 1.1. Geographical distribution of ATL patients in Japan.

et al[6] from these T cell lines. It is also important to mention the early NIH study of interleukin-2 (IL-2) which was started as T cell growth factor (TCGF) by Morgan et al[7] Using cultured T cells from the patient as the immunogen, the anti-Tac monoclonal antibody recognizing IL-2 receptor was obtained by Uchiyama et al.[8]

VIRUS DISCOVERY IN THE U.S. AND JAPAN

The causative agent for ATL was discovered in both the U.S. and Japan around 1980. Retrovirus HTLV-I had been isolated from the T cell lines established from CTCL. In Japan, Hinuma and

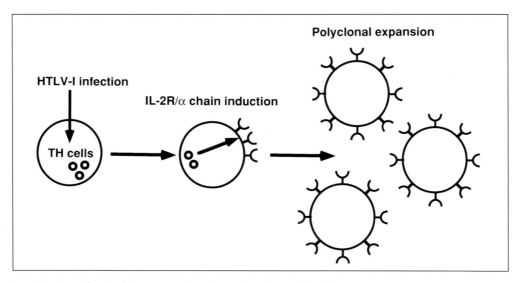

Fig. 1.2. Hypothesis of HTLV-I mediated transformation of T cells.

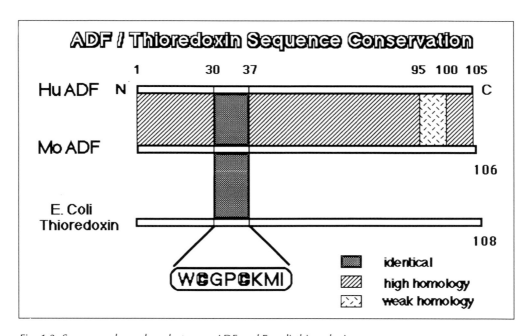

Fig. 1.3. Sequence homology between ADF and E. coli thioredoxin.

History 9

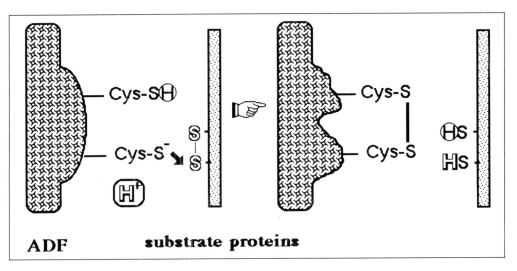

Fig. 1.4. Sulfhydryl reduction of proteins by ADF.

cooperating research groups found that the antigen was reactive with ATL patients' sera. The antigen, ATL-associated antigen (ATLA) later proved to be associated with the retrovirus, HTLV-I.[9]

Extensive clinical studies and studies on the molecular structure of the HTLV-I gene show that the viral infection resulted in the T cell activation. These studies showed that T cells express not only CD4 helper T cell marker antigen, but the Tac antigen which was later associated with the IL-2 receptor system α chain.[10]

MECHANISM OF ATL LEUKEMOGENESIS RELATED TO IL-2R DYSREGULATION

IL-2 AND OTHER CYTOKINE RECEPTOR SYSTEMS

During the early study of ATL cells, it was found that ATL cell lines constitutively expressed an activation marker Tac, which subsequently proved to be the IL-2 receptor. Today, the IL-2 receptor (IL-2R) is known to have a multichain structure in which the Tac antigen has been identified as the α chain of the receptor.[11-13]

The constitutive expression of the IL-2R α chain is considered to be one of the unique characteristics of ATL cells. The possible contribution of IL-2/IL-2R signaling in the leukemogenic process remains unknown. Despite fragmented evidence for the possible

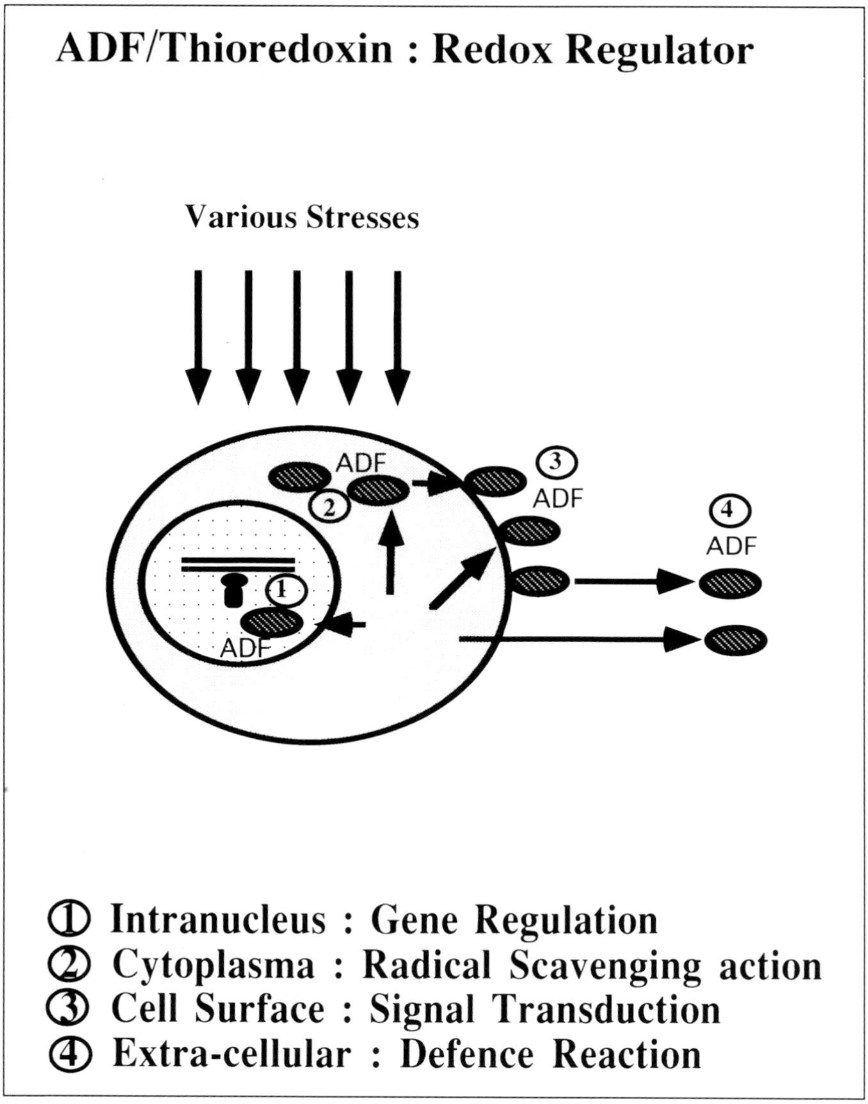
ADF/Thioredoxin : Redox Regulator
Various Stresses
ADF
2
3
ADF
4
ADF
1
ADF
① Intranucleus : Gene Regulation
② Cytoplasma : Radical Scavenging action
③ Cell Surface : Signal Transduction
④ Extra-cellular : Defence Reaction

Fig. 1.5. Variety in the site of action of ADF.

role of IL-2 in some steps of the leukemic process of ATL in vivo, this question has not yet been resolved (Fig. 1.2).[14-18]

One important question is whether HTLV-I$^+$ T cell lines established from ATL patients or HTLV-I$^+$ "healthy" carriers represent abnormal T cells which have expanded monoclonally in vivo. Indeed, an early study by Maeda et al in which they analyzed the T cell receptor α and β chain gene and integration site of HTLV-I

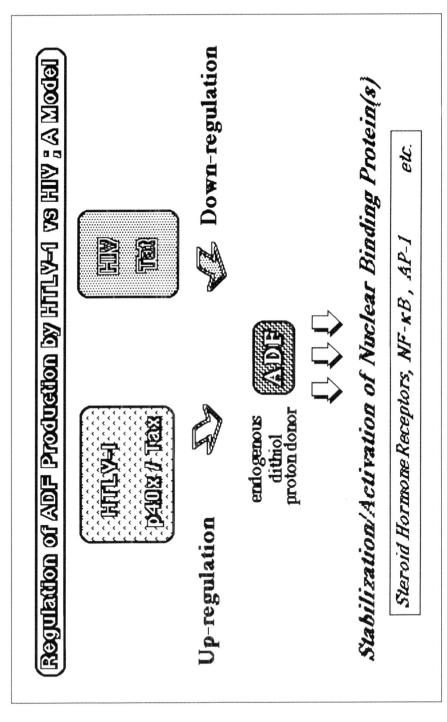

Fig. 1.6. Hypothesis of regulation of ADF production by HTLV-I and HIV.

provirus in the host genome, clearly demonstrated that not all of the HTLV-I⁺ T cell lines are leukemic in origin.[19] The mechanism of both in vivo and in vitro clonal expansion of HTLV-I-infected T cells and the possible contribution of cytokines, HTLV-I gene products as well as other cellular gene products, could aid in solving the leukemogenic process in ATL.

Redox Dysregulation in ATL

In the search for a possible mechanism of IL-2Rα/Tac overexpression, a soluble factor responsible for this receptor abnormality was discovered by Teshigawara et al.[20,21] The factor called ATL-derived factor (ADF) enhanced the IL-2Rα expression on some lymphocytes and lymphoid cell lines.[22,23] After gene cloning by Tagaya et al, ADF proved to be the human homologue of dithiol-related reducing enzyme thioredoxin, which has a variety of functions including disulfide reducto-oxidation, catalase-like radical scavenging, anti-apoptosis and co-cytokine activities (Figs. 1.3, 1.4).[24,25]

The intracellular function of ADF/thioredoxin (TRX) is to facilitate protein-protein and protein-nucleotide interactions such as NF-κB and AP-1 binding to DNA (Fig. 1.5).[26-28] Overproduction of ADF/TRX in HTLV-I-infected cells as well as the stress-inducibility of this protein have led to the assumption that dysregulation of stress responses, particularly against oxystress, is involved in the pathophysiology of a variety of viral infections including ATL and AIDS (Figs. 1.5, 1.6).

Relation of Other Disease Phenotypes

In the past ten years, additional clinical features of HTLV-I infection have been demonstrated. While the disease entity proposed as ATL actually contained a heterogeneous set of clinical features, clear evidence for the existence of the non-malignant disease condition in HTLV-I-infected individuals also emerged. The current paradigm on this issue is that these non-malignant diseases may be an independent disease phenotype of HTLV-I infection.

HTLV-I Associated Myelopathy (HAM)

Osame and others first described the association of a peculiar spastic disease frequent in the Kagoshima district on Kyushu Island and also of HTLV-I infection in that same district, where the incidence of ATL and the proportion of HTLV-I-infected individuals are higher than in other areas in Japan.[29] This disease, called HTLV-I associated myelopathy (HAM), later proved to be essentially identical to the tropical neurological disease, Tropical Spastic Paralysis (TSP), frequent in the Caribbean, where HTLV-I infection is also high.[30]

HTLV-I Related Arthropathy
Possible involvement of unknown viruses in autoimmune disorders has stimulated many investigators to identify HTLV-I-related virus gene integration in various patients with arthritis and other autoimmune manifestations. It is now claimed that in some of the rheumatoid arthritis (RA) patients there is an integration of HTLV-I in sinovial cells.[31] This disease entity is called HTLV-I-associated arthropathy (HAAP). The actual contribution of the virus in the pathophysiology of this disease has yet to be clarified, however.

The possible involvement of HTLV-I in some steps of the autoimmune disease process apparently enhanced the enthusiasm to find endogenous or exogenous HTLV-I-like viruses in such diseases.

HTLV-I-Related Uveitis
As is the case with HAM, there is a high incidence of uveitis and other ocular disorders in HTLV-I$^+$ populations in the Kyushu area. Although the direct relationship between the virus and these diseases is still unclear, the disorder is tentatively called HTLV-I-related uveitis.[32]

HTLV-I-Related Pulmonary Manifestations
There are reports claiming pulmonary manifestations of HTLV-I-related disorders. However, the actual relationship between these pulmonary pathological reports and ATL is still largely unknown.[33]

There is definite evidence showing that non-malignant diseases related to HTLV-I infection are present. The relationship between these diseases and ATL is important from both clinical and basic standpoints. Different medical approaches are needed for non-malignant and malignant diseases. The possible progression from non-malignant to malignant diseases is still partly speculative and must be critically elucidated.

Acquired Immunodeficiency Syndrome (AIDS)

In the early 1980s, an infectious disorder called AIDS was recognized among homosexual communities in the U.S. and other developed countries.

NIH investigators once hypothesized that HTLV-I or a closely associated virus was involved in AIDS. and there was actually a short period of apparent confusion about the retroviruses related to ATL and AIDS.[34] Subsequent studies in various countries, however, have disclosed that new viruses, not identical but distantly related to HTLV-I, are really the causative agents of AIDS.[35]

REFERENCES

1. Yodoi J, Takatsuki K, and Masuda T. Two cases of T-cell chronic lymphocytic leukemia in Japan. New Engl J Med 1974; 290:572-3.
2. Uchiyama T, Yodoi J, Sagawa K et al. Adult T cell leukemia. Clinical and hematologic features of sixteen cases. Blood 1977; 50:481-492.
3. Ishii K, Yodoi J, Hanaoka M et al. A hypotetraploid human T lymphoid cell line established by cell fusion. J Cell Physiol 1978; 94:93-8.
4. Miyoshi I, Kubonishi I, Yoshimoto S et al. Detection of type C virus particles in a cord leukocytes and human leukemic T-cells. Nature 1981; 296:770-1.
5. Gazdar AF, Carney DN, Bunn PA et al. Mitogen requirements for the in vitro propagation of cutaneous T-cell lymphomas. Blood 1980; 55:409-17.
6. Poiesz BZ, Ruscetti FW, Gazdar AF et al. Detection and isolation of type C retrovirus particles from fresh and cultured lymphocytes of a patient with cutaneous T cell lymphoma. Proc Natl Acad Sci USA 1980; 77:7415-9.
7. Morgan DA, Ruscetti FW and Gallo RC. Selective in vitro growth of T-lymphocytes from normal human bone marrows. Science. 1976; 193:1007.
8. Uchiyama T, Broder S, Waldmann TA. A monoclonal antibody (anti-Tac) reactive with activated and functionally mature human T cells. T. Production of anti-Tac monoclonal antibody and distribution of Tac$^+$ cells. J Immunol 1981; 126:1393-7.
9. Seiki M, Hattori S, Yoshida M. Human adult T-cell leukemia virus: Complete nucleotide sequence of the provirus genome integrated in leukemia cell DNA. Proc Natl Acad Sci 1983; 80:3618-22.
10. Leonard WJ, Depper JM, Uchiyama T et al. A monoclonal antibody that appears to recognize the receptor for human T-cell growth factor; partial characterization of the receptor. Nature 1982; 300:267-9.

11. Leonard WJ, Depper JM, Crabtree GR et al. Molecular cloning and expression of cDNAs for the human interleukin-2 receptor: Evidence for alternate mRNA splicing and the use of two polyadenylation sites. Nature 1984; 311:626-31.
12. Nikaido T, Shimizu A, Ishida N et al. Molecular cloning of cDNA encoding human interleukin 2 receptor. Nature 1984; 311;631-5.
13. Kondo S, Shimizu A, Maeda M et al. Expression of functional human interleukin-2 receptor in mouse T cells by cDNA transfection. Nature 1986; 320:75-7.
14. Hattori T, Uchiyama T, Toibana T et al. Surface phenotype of Japanese adult T-cell leukemia cells characterized by monoclonal antibodies. Blood 1981; 58:645-7.
15. Yodoi J, Uchiyama T, Maeda M. T-cell growth factor receptor in adult T-cell leukemia. Blood 1983; 62:509-510.
16. Tsudo M, Uchiyama T, Uchino H et al. Failure of regulation of Tac antigen/TCGF receptor on adult T cell leukemia cells by anti-Tac monoclonal antibody. Blood 1983; 61:1014-16.
17. Arima N, Daitoku Y, Ohgaki S et al. Autocrine growth of interleukin 2-producing leukemic cells in a patient with adult T cell leukemia. 1986; 68:779-782.
18. Maeda M, Arima N, Daitoku Y et al. Evidence for the interleukin-2 dependent expansion of leukemic cells in Adult T Cell Leukemia. Blood 1987; 70:1407-11.
19. Maeda M, Shimizu A, Ikuta K et al. Origin of HTLV-I$^+$ T-cell lines in adult T-cell leukemia: analysis on T-cell receptor gene rearrangement. J Exp Med 1985; 162:2169-2174.
20. Teshigawara K, Maeda M, Nishino K et al. Adult T leukemia cells produce a lymphkine that augments interleukin-2 receptor expression. J Mol Cell Immunol 1985; 2:17-26.
21. Okada M, Maeda M, Tagaya Y et al. TCGF(IL-2)-receptor inducing factor(s). II. Possible role of ATL-derived factor (ADF) on constitutive IL-2 receptor expression of HTLV-I$^+$ T cell lines. J Immunol 1985; 135:3995-4003.
22. Tagaya Y, Maeda Y, Mitsui A et al. ATL-derived factor (ADF), an IL-2 receptor/Tac inducer homologous to thioredoxin; Possible involvement of dithiol-reduction in the IL-2 receptor induction. EMBO J 1989; 8:757-764.
23. Tagaya Y, Masutani H, Nakamura H et al. Role of ATL-derived factor (ADF) in the normal and abnormal cellular activation. Involvement of dithiol related reduction. Mol Immunol 1990; 27:1279-1289.
24. Holmgren A. Thioredoxin catalyzes the reduction of insulin disulfides by dithiolthreitol and dihydrolipoamide. J Biol Chem 1979; 254:9627-9632.
25. Holmgren A. Thioredoxin. Ann Rev Biochem 1985; 54:237-271.
26. Hentze MW, Rouault TA, Harford JB et al. Oxidation-reduction

and molecular mechanism of a regulatory RNA-protein interaction. Science 1989; 244:357-9.
27. Abate C, Patel L, Rauscher FJIII et al. Redox regulation of Fos and Jun DNA-binding activity in vitro. Science 1990; 249:1157-61.
28. Schreck R, Rieber P, and Baeuerle PA. Reactive oxygen intermediates as apparently widely used messengers in the activation of the NB-κB transcription factor and HIV-1. EMBO J 1991; 10:2247-2258.
29. Osame M, Usuki K, Izumo S et al. HTLV-I associated myelopathy. A new clinical entity. Lancet 1985; 1:1031-2.
30. Gessain A, Barin F, Vermant JC et al. Antibodies to human T-lymphotropic virus type I in patients with tropical spastic paraparesis. Lancet 1985; ii:407-10.
31. Kitajima I, Yamamoto K, Sato K et al. Detection of HTLV-I proviral DNA and its gene expression in synovial cells in chronic inflammatory arthritis. J Clin Invest 1991; 88:1315-22.
32. Mochizuki M, Watanabe T, Yamaguchi K et al. HTLV-I uveitis: a distinct clinical entity caused by HTLV-I. Jpn J Cancer Res 1992; 83:236-9.
33. Sugimoto M, Nakashima H, Watanabe S et al. T lymphocytic alveolitis in HTLV-I-associated myelopathy. Lancet 1987; ii:1220.
34. Human T-cell leukemia/lymphoma Virus. Gallo RC, Essex ME, and Gross L. Eds. Cold Spring Harbor Lab. 1984.
35. Gallo RC, and Streicher HZ. AIDS Modern Concepts and Therapeutic Challenges. Broder S, ed. New York: Marcel Dekker Inc, 1987:1-21.

CHAPTER 2

CLINICAL FEATURES OF ATL AND OTHER HTLV-I-RELATED DISEASES

Takashi Uchiyama

ATL

INCIDENCE, ETIOLOGY AND CLASSIFICATION

ATL cases have been reported in Japan, the United Kingdom, the United States, the West Indies, Taiwan, South America, Central Africa and other countries.[1-8] The number of people infected with HTLV-I is estimated at about one million. Several hundred newly diagnosed ATL cases are reported annually in Japan, which has the highest ATL prevalence in the world.[9] The mean age of ATL patients at disease onset is 55 years and the ratio of male to female patients is 1.4 to 1.[9] HTLV-I is considered to be a causative agent of ATL based on the following: (1) patients with ATL almost always have serum antibodies against HTLV-I;[10-12] and (2) HTLV-I provirus is monoclonally integrated in leukemic cells of patients, which indicates that the proliferation and expansion of cells derived from a clone infected with HTLV-I results in the development of ATL.[13] Monoclonality of the leukemic cells from ATL patients has also been confirmed by analysis of the T cell receptor gene rearrangement pattern using Southern blot hybridization.

ATL is clinically classified into four types: acute, chronic, smoldering and lymphoma types (Table 2.1).[14-16] The acute type is

---

Adult T Cell Leukemia and Related Diseases, edited by Takashi Uchiyama and Junji Yodoi. © 1995 R.G. Landes Company.

Table 2.1. Diagnostic criteria for clinical subtypes of ATL

	smoldering	chronic	lymphoma	acute
Anti-HTLV-I antibody	+	+	+	+
Lymphocyte (x10^9/L)	<4	≥4^a	<4	
Abnormal T lymphocytes	≥5%	+^{b}	≤1%	+^{b}
Flower T lymphocytes	Occasionally	Occasionally	No	+
Serum LDH	≤1.5^N	≤2^N		
Corrected Ca (mmol/L)	<2.74	<2.74		
Histology-proven lymphadenopathy	No		+	
Tumor lesions				
Skin	c			
Lung	c			
Lymph node	No		Yes	
Liver	No			
Spleen	No			
Central nervous system	No	No		
Bone	No	No		
Ascites	No	No		
Pleural effusion	No	No		
Gastrointestinal tract	No	No		

N: Normal upper limit.
a: Accompanied by T lymphocytosis (3.5x10^9/L or more).
b: In case abnormal T lymphocytes are less than 5% in peripheral blood, histology-proven tumor lesion is required.
c: No essential qualification if other terms are fulfilled, but histology-proven malignant lesion is required incase abnormal T lymphocytes are less than 5% in peripheral blood.
Modified from Lymphona Study Group 1984-1987.[16]

characterized by acute presentation of symptoms such as fever, cough, dyspnea, abdominal pain and fullness, skin lesions, hypercalcemia, lymph node enlargement, hepatosplenomegaly, high white blood cell count, high serum levels of LDH, hyperbilirubinemia and the appearance of morphologically characteristic leukemic cells, and is usually accompanied by a rapid downhill course.[1,15,16] Chronic type ATL is characterized by milder symptoms and signs, a longer clinical course, and eventually leads to either an abrupt exacerbation of the disease, termed crisis, or in fatal complications such as pulmonary infections by microorganisms.[14-16] Patients with smoldering type ATL have a few leukemic cells in their peripheral blood and frequently present skin lesions such as papules, erythema and nodules,[14-17] but peripheral lymph node enlargement and

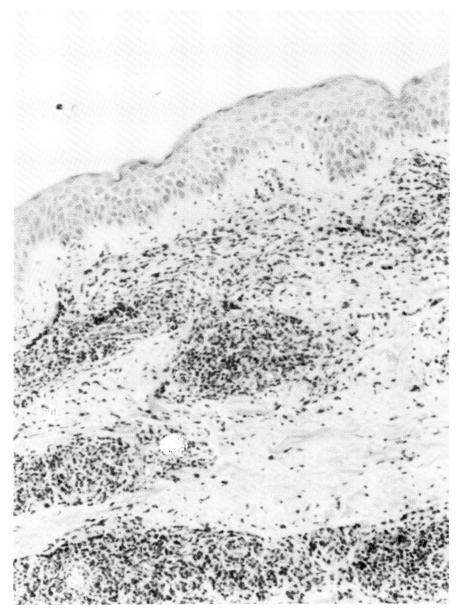

Fig. 2.1. Leukemic cell infiltration into skin in ATL. Leukemic cells predominantly infiltrate into dermis and subcutaneous tissue.

splenomegaly are minimal, and LDH serum level is either slightly elevated or normal. The clinical picture of lymphoma type ATL mainly consists of lymph node enlargement.[14-16] The Lymphoma Study Group

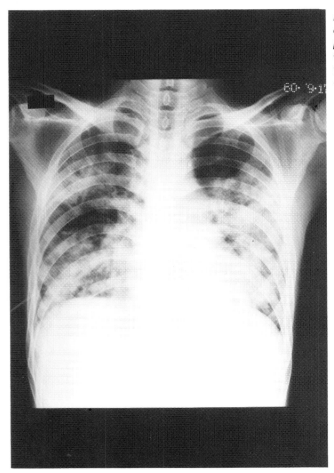

Fig. 2.2. Chest X-ray of of ATL patient with pneumonia due to Cytomegalovirus, Aspergillus and Pneumocystis carinii.

proposed the following criteria for diagnosis of ATL types:[16]
1. Smoldering type: peripheral blood lymphocyte count less than 4×10^9/L; 5% or more abnormal T lymphocytes; occasional appearance of flower-like T cells; serum LDH level up to 1.5 times the normal upper limit; no hypercalcemia; no histologically proven lymphadenopathy; no tumor lesions in any of lymph nodes, liver, spleen, central nervous system (CNS), bone and gastrointestinal (GI) tract; no pleural effusion; no ascites; occasional skin and/or lung lesions by tumors.
2. Chronic type: peripheral blood lymphocyte count more than 4×10^9/L; more than 5% abnormal T lymphocytes

or histologically proven tumor lesion; occasional flower-like T cells; serum LDH level up to twice the normal upper limit; no hypercalcemia; no involvement of CNS, bone and GI tract; no pleural effusion, no ascites. Patients may have lymphadenopathy, hepatomegaly, splenomegaly, skin and pulmonary lesions.
3. Lymphoma type: histologically proven lymphadenopathy; no leukemic manifestation (lymphocyte count less than 4×10^9/L and 1% or less abnormal T lymphocytes); possible involvement of skin, lung, lymph node, liver, spleen, CNS, bone or GI tract.
4. Acute type, which is defined as remaining ATL which is not classified as smoldering, chronic or lymphoma type. Patients with acute type ATL usually have both leukemic manifestations and tumor lesions such as lymphadenopathy and extranodal lesions. The relative percentage of each type of ATL case are roughly: 55% acute type; 20% lymphoma type; 20% chronic type; and 5% smoldering type.

CLINICAL FEATURES

Symptoms and Signs

Symptoms at onset of the disease include general malaise, fever, cough, dyspnea, abdominal fullness, abdominal pain, jaundice, superficial lymph node swelling, skin manifestations such as papules, nodules, nodes, erythema and erythroderma, thirst and drowsiness. Thirst, drowsiness and constipation are due to the hypercalcemia which is often encountered in the disease. Physical findings frequently detectable at diagnosis include lymphadenopathy, hepatomegaly, splenomegaly, various skin lesions, jaundice and ascites.

Skin lesions include papules, nodules, nodes, tumors, erythema and erythroderma. There are no skin lesions which are pathognomonic of ATL. Histological examination of affected skin usually shows infiltration of leukemic cells into dermis and subcutaneous tissue (Fig. 2.1). Leukemic cell infiltration into epidermal tissue, which is almost always observed in Sézary syndrome,[18] is also occasionally detectable in ATL.

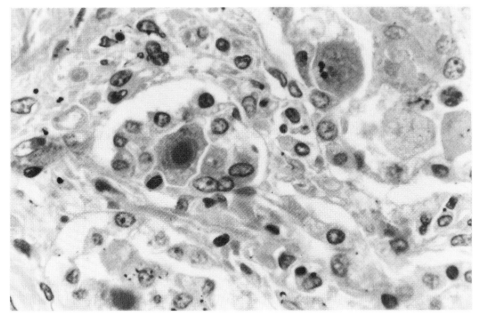

Fig.2.3A.

Fig. 2.3. Histopathological findings of the lung of an ATL patient with pneumonia due to (A) Cytomegalovirus, (B) Aspergillus, and (C) Pneumocystis carinii.

Pulmonary changes, which are often detectable by chest X-ray, are predominantly composed of either neoplastic cell infiltration into pulmonary tissue or pulmonary infections by bacteria, fungi and/or viruses (Fig. 2.2). The most commonly observed pulmonary infections are pneumonias due to Cytomegalovirus, Aspergillus and Pneumocystis carinii as shown in Figures 2.3A, 2.3B and 2.3C, respectively. Figure 2.2 shows a chest X-ray film of a patient who developed pneumonia due to Cytomegalovirus, Aspergillus and Pneumocystis carinii (Figs. 2.3A-2.3C).

Hepatomegaly and abnormal findings such as hyperbilirubinemia and elevated levels of serum GPT and GOT are usually due to infiltration of leukemic cells into the liver. Leukemic cell infiltration and/or tumor formation in the gastrointestinal tract are also sometimes observed.[19] Central nervous system involvement in ATL, however, is uncommon.[20]

Laboratory Findings

Peripheral blood lymphocyte count is high, especially in acute type ATL. Marked leukocytosis of more than $30 \times 10^9/L$ is observed

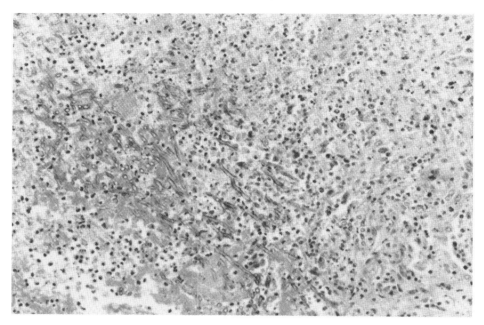

Fig.2.3B.

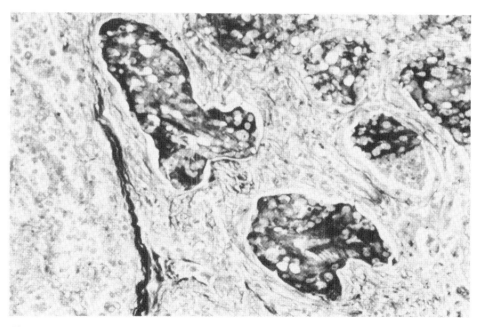

Fig.2.3C.

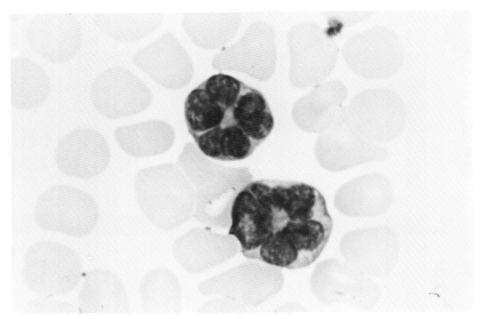

Fig. 2.4. Typical leukemic cells from ATL patients which have lobulated or deeply indented nuclei.

in about 40% of acute type and 20% of chronic type ATL cases. Peripheral blood smears usually show the characteristic leukemic cells which have a large deeply-indented or lobulated nuclei with clear nucleoli in basophilic cytoplasms.[1] Typical leukemic cells sometimes bear a flower-like configuration (Fig. 2.4). The mechanism of the development of such morphologically peculiar cells remains unknown. Another characteristic finding in peripheral blood smears is the pleomorphism of leukemic cells. Cells with variable sizes and different morphologic changes may be detected in the same blood smears. The relative proportion of the morphologically characteristic cells is usually high in acute type ATL and low in other types of ATL. Patients with smoldering type or lymphoma type ATL do not have leukocytosis, but neutrophilia (30% of all cases) or eosinophilia are not uncommon. Anemia and thrombocytopenia are not common, although severe anemia and thrombocytopenia are reported in rare cases.[21]

Bone marrow aspiration reveals an increase in the proportion of lymphoid cells, including characteristic leukemic cells. However, this increase is not marked in a small proportion of acute type and in a majority of chronic and lymphoma type ATL.

Histopathological examination of enlarged lymph node biopsy specimens showed destruction of normal structure as evidenced by burned-out lymph follicles, uncertain peripheral sinus, and predominant distribution of atypical lymphoid cells.[22,23] There is evidence of diffuse proliferation of atypical lymphoid cells which vary in size and shape and have irregular or convoluted nuclear contours, stippled chromatin patterns and amphophilic cytoplasm. Giant cells with cerebriform or Reed-Sternberg type nuclei and mitoses are also occasionally observed. These gian T cells are usually classified into diffuse/pleomorphic type, diffuse/mixed type or diffuse/large cell type, according to the classification criteria proposed by Lymphoma Study Group (LSG).[24]

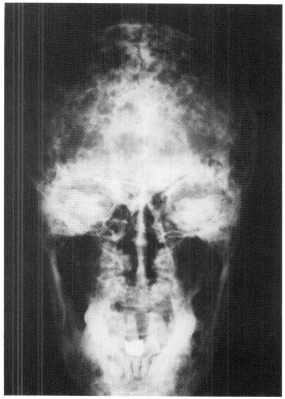

Fig. 2.5. Marked osteolytic lesions of the skull in an ATL patient with hypercalcemia.

Serum lactate dehydrogenase (LDH) level is usually moderately to markedly elevated in both acute and lymphoma type ATL. Serum levels of the soluble form of IL-2 receptor α chain are markedly (usually more than 10 to 100 times higher than normal except for smoldering type ATL) high and serve as a very useful indicator of the whole mass of neoplastic cells in the body.[25,26] Serum levels of total bilirubin, transaminases (e.g. GOT, GPT) and alkaline phosphatase are elevated when leukemic cells infiltrate into the liver. Liver dysfunction and hypoalbuminemia (less than 40 g/L) are detectable in about 60% and 74% of acute type

Table 2.2. Prognosis of ATL

	Mean Survival Time (Months)	Survival Rate at	
		2 years(%)	4 years(%)
Acute Type	6.2	16.7	5.0
Lymphoma Type	10.2	21.3	5.7
Chronic Type	24.3	52.4	26.9
Smoldering Type	–	77.7	62.8
All Cases	9.0	26.8	10.3

(Lymphona Study Group 1984-1987)[16]
Reproduced with permission from Shimoyama et al, Br J Haematol 1991; 79:430.

ATL, cases respectively. Renal dysfunction, as evidenced by elevated levels of blood urea nitrogen and serum creatinine, is observed in about 30% of acute type, 15% of lymphoma type and 10% of chronic type ATL cases. Hypercalcemia (corrected serum level more than 2.74 mmol/L) is detectable in 50% of acute type and 15% of lymphoma type ATL cases.[27,28]

COMPLICATIONS AND PROGNOSIS

ATL patients often suffer from serious complications such as infections by bacteria, fungi, protozoa and viruses. Common infections are bacterial pneumonia, Pneumocystis carinii pneumonia, Aspergillus or Candida pneumonia, Strongyloidiasis and Cytomegalovirus pneumonia. Strongyloidiasis has often been observed in HTLV-I-infected carriers and ATL patients in Okinawa.[29] Approximately half of ATL patients die of severe infections. Among those infections which are direct causes of death in ATL patients, the most frequent (80%) is pneumonia, followed by sepsis. Bacterial, fungal, viral (including interstitial pneumonitis) and protozoal pneumonia are 50%, 10%, 40% and 5%, respectively, of all the pneumonias.[16] Frequent infectious complications probably result from the immunodeficient state of ATL patients. Perturbations of the immune system, as indicated by the abnormal cytokine network and abnormal antigen recognition and signal transduction through the T cell receptor/CD3 complex, which may be induced by the dysfunction of HTLV-I-infected T cells, could be one of the mechanisms underlying immunodeficiency or the immunosuppressed state in ATL.

Hypercalcemia, a serious and bothersome complication, is observed in 40 to 50% of all patients throughout the entire clinical course.[16] It has been reported that cytokines including IL-1α, IL-1β, TNF-β and TGF-β are abnormally produced in HTLV-I-infected cell line cells and/or fresh leukemic cells[30-36] and may be involved in hypercalcemia. It is probably predominantly due to excessive production of parathyroid hormone related protein (PTHrP), the transcription of which is enhanced by Tax of HTLV-I.[37-39] Since PTHrP mobilizes calcium ions from bones, marked osteolytic lesions may be demonstrated by roentgenological examination in some cases (Fig. 2.5). Osteolytic lesions are observed not only in the skull, vertebrae and costal ribs, but also in peripheral bones of extremities, in contrast to those observed in multiple myeloma. Histopathological examination of affected bone shows proliferation of osteoclasts and bone resorption. Very few cases of spontaneous regression or remission of disease have been reported.[40,41] An analysis of such cases may provide clues on the nature of neoplastic cell growth in vivo and the mechanism of disease occurrence and progression. Disease prognosis is generally very poor in ATL cases (Table 2.2). The median survival time based on 818 ATL cases in a 1984-1987 LSG study[16] was 9 months, and the survival rate at 2 and 4 years was 26.8% and 10.3%, respectively. Acute and lymphoma type ATL had a very poor prognosis, while the prognosis of chronic and smoldering types was less grave. Median survival time was 6.2 months for the acute type, 10.2 months for the lymphoma type, 24.3 months for the chronic type, and has not yet been calculated for the smoldering type.[16] The 2 and 4 year survival rates were 16.7% and 5.0% for the acute type, 21.3% and 5.7% for the lymphoma type, 52.4% and 26.9% for the chronic type, and 77.7% and 62.8% for the smoldering type, respectively[16] (Table 2.2).

TREATMENT

Combination chemotherapy has been the major treatment employed for ATL. This treatment can be roughly classified into three or four types of therapy: old type; the first generation; and second or third generation combination chemotherapy. The old type combination chemotherapy consists of COP (cyclophosphamide, vincristine, prednisolone), VEMP (vincristine, cyclophosphamide,

6-mercaptopurine, prednisolone), and others. The first generation combination chemotherapy consists of VEPA (vincristine, cyclophosphamide, prednisolone, adriamycin),[42,43] VEPA-M (VEPA + methotrexate),[20,44] CHOP (cyclophosphamide, adriamycin, vincristine, prednisolone)[20] and others. The second or third generation combination chemotherapy consists of LSG4 (alternating therapy using three different regimens, VEPA-B + M-FEPA + VEPP-B),[45,46] MACOP-B (methotrexate, adriamycin, cyclophosphamide, vincristine, prednisolone, bleomycin) and others. Complete remission rates for the old type, first generation and second or third generation combination chemotherapies are 11%, 21% and 41%, respectively. These results indicate improvement in the induction of remission through combinations of intensified doses of anti-tumor drugs.[20,42-46] Nevertheless, disease-free survival time and overall survival time have not been improved by such new generation combination chemotherapy. Other treatment trials such as bone marrow transplantation,[47-48] low dose total body irradiation,[49] leukopheresis, administration of interferon β, interferon γ,[50] and IL-2 were found to be unsuccessful. Deoxycoformycin (DCF), adenosine deaminase inhibitor was found to be effective in about a half of cases of Sézary syndrome and T-prolymphocytic leukemia.[51] DCF was recently tested in ATL patients and showed a response rate of about 30%.[52] Another new agent, bis(2,6-dioxopiperazine), an inhibitor of topoisomerase II, was also clinically tested for treatment of ATL. A total of 10 (two cases with complete and eight cases with partial remission) of 23 cases tested were responsive to administration of this new drug. The duration of remission was not long, however, and ranged from 43 to 374 days.[53]

Basic studies for IL-2 receptor targeted therapy have been performed. Its clinical application has also been tested in ATL patients in the U.S. The administration of mouse monoclonal antibody, anti-Tac which reacts with human IL-2 receptor α chain resulted in clinical improvement in 7 of 19 patients examined.[54] Anti-Tac antibody conjugated with ^{90}Y was also effective in 10 of 15 patients with ATL.[55] Anti-Tac antibody conjugated with ricin A chain,[56,57] Pseudomonas exotoxin,[58] or ^{212}Bi[55] and the fusion protein composed of IL-2 and Diphtheria toxin[59,60] or Pseudomonas exotoxin[61] also selectively kill HTLV-I-infected T cell line cells in vitro which express a large number of IL-2 receptor α chains.

OTHER HTLV-I-RELATED DISEASES

HTLV-I-Associated Myelopathy (HAM)/Tropical Spastic Paraparesis (TSP)

In 1985, Gessain et al first demonstrated antibodies against HTLV-I in 59% of a group of patients with tropical spastic paraparesis (TSP) whose clinical features and pathological aspects had been studied extensively.[62-65] Meanwhile, Igata, Osame and colleagues independently noted myelopathic patients in southern Japan with peculiar clinical features and subsequently associated these myelopathies with HTLV-I infection[66-68] which they called HTLV-I-associated myelopathy (HAM). Subsequent comparative studies revealed that TSP and HAM were identical diseases, and it was thus proposed to call it HAM/TSP.[69] HAM/TSP cases were reported in Japan, the U.S., Jamaica, Martinique, Colombia, Chile, Brazil, England, Seychelles, Zaire, South Africa, and other countries.[70-72] The incidence of HAM/TSP among HTLV-I-infected people was estimated to be about 3×10^{-5} cases/year.[68]

The mean age of disease onset is 43 years and the male to female ratio of the patients is 1:2.9 in Japan.[68] The period from HTLV-I infection to disease occurrence may range from months to decades, which is still shorter than the latent period for ATL. A quarter of Japanese HAM/TSP cases have a history of blood transfusion[68] although recently, the number of transfusion-associated HAM cases have rapidly declined due to the use of screening tests to detect HTLV-I antibodies. The majority of the remaining cases may be associated with mother-to-baby transmission of HTLV-I.

The main neurological manifestations of HAM/TSP are as follows: (1) chronic spastic paraparesis, which usually progresses slowly, but sometimes remains static after initial progression; (2) weakness of lower limbs, which is more marked proximally; (3) bladder disturbance, constipation and impotence or decreased libido; (4) sensory symptoms such as tingling, pins and needles, burning, etc. which are more prominent than objective physical signs; (5) low lumbar pain with radiation to the legs; (6) impaired vibration sense; (7) hyperreflexia of the lower limbs, often with clonus and Babinski's sign; (8) hyperreflexia of upper limbs, often with positive Hoffmann's and Trömner signs; and (9) exaggerated jaw jerk.

Table 2.3. HTLV-I-associated diseases

1. Adult T cell leukemia (ATL)
2. HTLV-I-associated myelopathy (HAM)/Tropical spastic paraparesis (TSP)
3. HTLV-I-associated arthropathy (HAAP)
4. HTLV-I-associated uveitis (HAU)
5. HTLV-I-associated bronchopneumonopathy (HAB)
   HTLV-I-associated bronchoalveolar disorder (HABA)
6. Polymyositis
7. Infectious dermatitis

In addition, laboratory examinations revealed: (1) the presence of HTLV-I antibodies in the serum and cerebrospinal fluid (CSF); (2) a mild lymphocyte pleocytosis in the CSF; (3) lymphocytes with lobulated nuclei in blood and/or CSF; and (4) a mild to moderate increase of protein in the CSF. The titer of HTLV-I serum antibodies is usually higher in HAM/TSP than in ATL. Increases in the number of circulating activated T cells,[73] serum immunoglobulin levels,[72] spontaneous ^{3}H-TdR uptake by peripheral blood mononuclear cells[74,75] and replication of HTLV-I[75] have also been reported. The mechanisms involved in the development of HAM/TSP remain unknown. Immune system disturbances triggered by HTLV-I, which may be affected by genetic factors as indicated by preferred expression of HLA genotypes in patients, may play a key role in disease development.[77] The administration of prednisone usually provides transient improvement in early stages of the disease. Other clinical trials such as intrathecal administration of[78,79] hydrocortisone, interferon α and azathiopurine have also proved effective in some cases. The prognosis of HAM/TSP is usually not poor. The mean morbidity period of HAM/TSP patients is about 10 years, the major causes of death being complications like infections and cancers.

HTLV-I-Associated Arthropathy (HAAP)

A complication of polyarthritis was occasionally observed in HAM/TSP and rarely in ATL.[80,81] Thereafter, HTLV-I-associated arthropathy (HAAP) was proposed as a new disease entity and defined as an arthropathy developing in HTLV-I-infected individuals.[82] A summary of clinical characteristics of patients with HAAP follows:[80-82] (1) patients are most commonly found in southwestern

Japan, which is an endemic area of HTLV-I; (2) the age of HAAP patients is generally higher than the age of rheumatoid arthritis (RA) patients; (3) big rather than small joints, such as shoulders, knees and wrists are commonly affected; (4) various joint lesions are detectable; and (5) extra-articular symptoms such as bronchitis and myalgia are common. HTLV-I antibodies were detected in synovial fluids of affected joints and HTLV-I proviral DNA was demonstrable by a polymerase chain reaction (PCR), both in synovial tissue and in synovial fluid lymphocytes[83,84] of mice transgenic for HTLV-I.[85] Swelling of the ankle with redness and/or swelling of the footpad near the ankle were also observed in these mice. Histopathological examination of affected joints revealed erosion of synovial bones and cartilage, pannus-like granulation tissue composed of fibroblasts and small vessels with infiltration of lymphocytes, neutrophils and macrophages, and synovial proliferation associated with stratification of synovial lining cells, which closely resembled changes observed in human RA cases. The mRNA of Tax was markedly expressed at the joints. Based on the body of evidence, it is likely that HTLV-I is involved in chronic arthritis in humans. It is possible that some cytokines which induce inflammation and/or bone destruction are abnormally produced by transacting transcriptional activity of Tax of HTLV-I, or that Tax may induce production of factors which induce the proliferation of synovial cells. Alternatively, immunologic disturbances, including autoimmunity which may be triggered by HTLV-I infection, may be involved in the development of HAAP.

HTLV-I Uveitis (HAU)

Uveitis is defined as an inflammatory lesion of iris, ciliary body and choroid, and adjacent tissues such as the vitreous body, retina and optic nerve are often involved. Various etiologic agents and mechanisms have been reported for uveitis.[86] Mochizuki et al noted that the relative proportion of uveitis without defined etiology was higher in Kurume and Miyakonojo, located in Kyushu, than that in Tokyo. They examined HTLV-I antibody in the sera of uveitis patients without defined etiology and found a high antibody positive rate (21.8%) in Miyakonojo.[87] Further studies demonstrated HTLV-I proviral DNA by PCR in infiltrating cells in the aqueous humor of anterior chamber of the affected eyes.[88] Based on seroepidemiological, clinical and virological data obtained, they suggested that HTLV-I was associated with a certain type of uveitis and proposed HTLV-I uveitis as a new clinical entity.[87-89] The clinical features they reported follow: (1) blurred vision or myodesopsia with acute or subacute onset as a chief complaint; (2) visual acuity which is relatively preserved in most cases; (3) frequently observed ocular findings such as iritis (97%), vitreous opacities (93%), retinal vasculitis (66%) and retinal exudates and hemorrhages (19%); and (4) mean patient age of 44 in men and 49 in women. HTLV-I uveitis usually responds well to ocular or oral administration of corticosteroids, although disease recurrence is observed in half of all cases.

OTHER HTLV-I-RELATED DISEASES

Other diseases which may be associated with HTLV-I infection include polymyositis,[90] chronic respiratory diseases,[91,92] lymphadenitis,[93] and dermatitis,[94] although it remains to be established that these diseases are distinct clinical entities.

REFERENCES

1. Uchiyama T, Yodoi J, Sagawa K et al. Adult T-cell leukemia: Clinical and hematologic features of 16 cases. Blood 1977; 50:481-92.
2. Catovsky D, Greaves MF, Rose M et al. Adult T-cell lymphoma-leukemia in blacks from the West Indies. Lancet 1982; i:639-42.
3. Bunn PA, Schechter GP, Jaffe ES et al. Clinical course of retrovirus-associated adult T-cell lymphoma in the United States. N Engl J Med 1983; 309:257-64.
4. Blattner WA, Gibbs WN, Saxinger C et al. Human T-cell leukemia/lymphoma virus-associated lymphoreticular neoplasia in Jamaica. Lancet 1983; ii:61-4.
5. Chen P, Chiu C, Chiou T et al. Adult T cell leukemia. First case reported in Taiwan. Acta Haematol Jpn 1985; 48:1035-41.
6. Pombo de Oliveira MS, Gollner AM, Serpa AMJ et al. Evidence of high prevalence HTLV-I positive in lymphoproliferative disorders (Brazil). Proc. of the 5th International Conference of Human Retrovirology; HTLV. 1992:2.
7. Gioseffi ON, Fantl D, Nucifora E et al. Adult T leukemia-lymphoma (ATLL)/HTLV-I in Argentina: Report of the three first cases. La Revista de Investigacion Clinica Supp to 1994:321.
8. Williams OK, Saxinger C, Junaid A et al. HTLV-associated lymphoproliferative disease: A report of two cases in Nigeria. Br Med J 1984; 288:1495-6.

9. Tajima K and The T- and B-cell Malignancy Study Group. The fourth nationwide study of adult T-cell leukemia/lymphoma (ATL in Japan): Estimates of risk of ATL and its geographical and clinical features. Int J Cancer 1990; 45:237-43.
10. Hinuma Y, Nagata K, Hanaoka M et al. Antigen in an ATL cell line and detection of antibodies to the antigen in human sera. Proc Natl Acad Sci USA 1981; 78:6476-80.
11. Hinuma Y, Komoda H, Chosa T et al. Antibodies to adult T-cell leukemia virus-associated antigen (ATLA) in sera from patients with ATL and controls in Japan: a nation-wide sero-epidemiologic study. Int J Cancer 1982; 29:631-5.
12. Blattner WA, Kalyanaraman VS, Robert-Guroff M et al. The human type-C retrovirus HTLV in blacks from the Caribbean region and relationship to adult T-cell leukemia/lymphoma. Int J Cancer 1982; 30:257-64.
13. Yoshida M, Seiki M, Yamaguchi K et al. Monoclonal integration of human T-cell leukemia provirus in all primary tumors of adult T-cell leukemia suggests causative role of human T-cell leukemia virus in disease. Proc Natl Acad Sci USA 1984; 81:2534-7.
14. Kawano F, Yamaguchi K, Nishimura H et al. Variation in the clinical courses of adult T-cell leukemia. Cancer 1985; 55:851-6.
15. Takatsuki K, Yamaguchi K, Kawano F et al. Clinical diversity in adult T-cell leukemia/lymphoma. Cancer Res 1985; 45(Supp): 4644-5.
16. Shimoyama M and the members of The Lymphoma Study Group (1984-87). Diagnostic criteria and classification of clinical subtypes of adult T-cell leukaemia-lymphoma. A report from the Lymphoma Study Group (1984-87). Br J Haematol 1991; 79:428-37.
17. Yamaguchi K, Nishimura H, Kohrogi H et al. A proposal for smoldering adult T-cell leukemia: a clinicopathologic study of five cases. Blood 1983; 62:758-66.
18. Lutzner M, Edelson R, Shein P et al. Cutaneous T-cell lymphomas: The Sezary syndrome, mycosis fungoides and related disorders. Ann Intern Med 1975; 43:534-52.
19. Hattori T, Asou N, Suzushima H et al. Leukemia of novel gastrointestinal T-lymphocyte population infected with HTLV-I. Lancet 1991; 337:76-7.
20. Shimoyama M, Ohta K, Kikuchi M et al. (For the Lymphoma Study Group (1984-1987)). Major prognostic factors of patients with advanced T cell lymphoma/leukemia. J Clin Oncol 1988; 6:1088-97.
21. Katsuno M, Uchida E, Gotoh K et al. Adult T cell leukemia presenting with pancytopenia followed by diabetes insipidus. Jap J Clin Hematol 1987; 28:730-7.
22. Hanaoka M, Sasaki M, Matsumoto H et al. Adult T cell leukemia. Histological classification and characteristics. Acta Pathol Jpn 1979;

29:723-38.
23. Kikuchi M, Mitsui T, Matsui N et al. T cell malignancies in adults. Histopathological studies of lymph nodes in 110 patients. Jpn J Clin Oncol 1979; 9(Suppl):407-22.
24. Suchi T, Tajima K, Namba K et al. Some problems on the histological diagnosis of non-Hodgkin's malignant lymphoma. A proposal of a new type. Acta Pathol Jpn 1979; 29:755-66.
25. Marcon L, Rubin LA, Kurman CC et al. Elevated serum levels of soluble Tac peptide in adult T-cell leukemia: Correlation with clinical status during chemotherapy. Ann Intern Med 1988; 15:274-9.
26. Motoi T, Uchiyama T, Uchino H et al. Serum soluble interleukin-2 receptor in patients with adult T-cell leukemia and human T-cell leukemia/lymphoma virus type-I seropositive healthy carriers. Jpn J Cancer Res(Gann) 1988; 79:593-9.
27. Grossman B, Schechter GP, Horton J et al. Hypercalcemia associated with T-cell lymphoma-leukemia. Am J Clin Pathol 1981; 75:149-55.
28. Kinoshita K, Kamihira S, Ikeda S et al. Clinical, hematologic, and pathologic features of leukemic T-cell lymphoma. Cancer 1982; 50:1554-62.
29. Nakada K et al. High incidence of HTLV antibody in carriers of strongyloides stercoralis. Lancet 1984; :633.
30. Wano Y, Hattori T, Matsuoka M et al. Interleukin 1 gene expression in adult T cell leukemia. J Clin Invest 1987; 80:911-6.
31. Kodaka T, Uchiyama T, Umadome H et al. Expression of cytokine mRNA in leukemic cells from adult T cell leukemia patients. Jpn J Cancer Res 1989; 80:531-6.
32. Noma T, Nakakubo H, Sugita M et al. Expression of different combinations of interleukins by human T cell leukemic cell lines that are clonally related. J Exp Med 1989; 169:1853-8.
33. Tschachler E, Robert-Guroff M, Gallo RC et al. Human T-lymphotropic virus I-infected T cells constitutively express lymphotoxin in vitro. Blood 1989; 73:194-201.
34. Tschachler E, Bohnlein E, Felzmann S et al. Human T-lymphotropic virus type I tax regulates the expression of the human lymphotoxin gene. Blood 1993; 81:95-100.
35. Niitsu Y, Urushizaki Y, Koshida Y et al. Expression of TGF-β gene in adult T cell leukemia. Blood 1988; 71:263-6.
36. Kim SJ, Kehrl JH, Burton J et al. Transactivation of the transforming growth factor β 1 (TGF-β 1) gene by human T lymphotropic virus type 1 tax: a potential mechanism for the increased production of TGF-β 1 in adult T cell leukemia. J Exp Med 1990; 172:121-9.
37. Honda S, Yamaguchi K, Miyake Y et al. Production of parathyroid hormone-related protein in adult T-cell leukemia cells. Jpn J Cancer Res 1988; 79;1264-8.

38. Motokura T, Fukumoto S, Takahashi S et al. Expression of parathyroid hormone-related protein in a human T lymphotropic virus type I infected T cell line in culture. Biochem Biophys Res Commun 1988; 154:1182-8.
39. Watanabe T, Yamaguchi K, Takatsuki K et al. Constitutive expression of parathyroid hormone-related protein gene in human T cell leukemia virus type I (HTLV-I) carriers and adult T cell leukemia patients that can be trans-activated by HTLV-I tax gene. J Exp Med 1990; 172:759-65.
40. Murakawa M, Shibuya T, Teshima T et al. Spontaneous remission from acute exacerbation of chronic adult T cell leukemia. Blut 1990; 61:346-9.
41. Shimamoto Y, Funai N, Suga K et al. Spontaneous regression in adult T-cell leukemia/lymphoma. Cancer 1993; 72:735-40.
42. Lymphoma Study Group: Shimoyama M, Yunoki K, Ichimaru M et al. Combination chemotherapy with vincristine, cyclophosphamide (Endoxan), prednisolone and adriamycin (VEPA) in advanced adult non-Hodgkin's lymphoid malignancies: Relation between T-cell or non-T-cell phenotype and response. Jpn J Clin Oncol 1979; 9(Supp):397-406.
43. Lymphoma Study Group (1978-80): Shimoyama M, Ichimaru M, Yunoki K et al. Final results of cooperative study of VEPA (vincristine, cyclophosphamide (Endoxan), prednisolone and adriamycin) therapy in advanced adult non-Hodgkin's lymphoma: Relation between T- or B-cell phenotype and response. Jpn J Clin Oncol 1982; 12:227-38.
44. Shimoyama M, Ota K, Kikuchi M et al.(For the Lymphoma Study Group (1981-1983)) Chemotherapeutic results and prognostic factors of patients with advanced non-Hodgkin's lymphoma treated with VEPA or VEPA-M. J Clin Oncol 1988; 6:128-41.
45. Minato K, Araki K, Hanada S et al.(Lymphoma Study Group (LSG)) An interim report of LSG4 treatment for advanced peripheral T-cell lymphoma. Int J Hematol 1991; 54(Supp 1):179.
46. Shimoyama M. Treatment of patients with adult T-cell leukemia/lymphoma: overview. In: Takatsuki K, Hinuma Y, Yoshida M ed. Advances in Adult T-cell Leukemia and HTLV-I Research. Japan Scientific Societies Press, Tokyo: 1992; 43-56.
47. Asou N, Sakai K, Yamaguchi K et al. Autologous bone marrow transplantation in a patient with lymphoma type adult T cell leukemia. Rinshou Ketsueki 1985; 26:229-33.
48. Sobue R, Yamauchi M, Miyamura K et al. Treatment of adult T cell leukemia with mega-dose cyclophosphamide and total body irradiation followed by allegeneic bone marrow transplantation. Bone Marrow Transplant 1987; 2:441-4.
49. Tamura K, Okayama A, Koga K et al. Total body irradiation as a primary treatment for adult T-cell leukemia. Jpn J Clin Oncol 1983;

13(Supp):313-24.
50. Tamura K, Makino S, Araki Y et al. Recombinant interferon β and γ in the treatment of adult T cell leukemia. Cancer 1987; 59:1059-62.
51. Dearden C, Matutes E, Catovsky D. Deoxycoformycin in the treatment of mature T-cell leukaemias. Br J Cancer 1991; 64:903-6.
52. Cooperative DCF Study Group: Tobinai K, Shimoyama M. Phase II study of YK-176 (2'-deoxycoformycin). Proc Jpn Assoc Cancer Res 1990; 49:376.
53. Ohno R, Masaoka T, Shirakawa S et al. Treatment of adult T-cell leukemia/lymphoma with MST-16, a new oral antitumor drug and a derivative of bis(2,6-dioxopiperazine). The MST-16 Study Group. Cancer 1993; 71:2217-21.
54. Waldmann TA, White JD, Goldman CK et al. The interleukin-2 receptor—a target for monoclonal antibody treatment of human T-cell lymphotrophic virus I-induced adult T-cell leukemia. Blood 1993; 82:1701-12.
55. Waldmann TA, Ira HP, Otta A et al. The multichain interleukin-2 receptor: a target for immunotherapy. Ann Intern Med 1992; 116:148-60.
56. Kronke M, Depper JM, Leonard WJ et al. Adult T cell leukemia: a potential target for ricin A chain immunotoxins. Blood 1985; 65:1416-21.
57. Uchiyama T, Kondo A, Kiyokawa T et al. Interleukin 2 receptor-targeted therapy of ATL. In: Ogura T and Takaku F ed. Regulation of Human Cancers by Cytokines: Present Status and Future Prospects. Tokyo: Japan Scientific Societies Press, 1993:125-33.
58. Kreitman JR, Chaudhary KV, Waldmann T et al. The recombinant immunotoxin anti-Tac(Fv)-Pseudomonas exotoxin 40 is cytotoxic toward peripheral blood malignant cells from patients with adult T-cell leukemia. Proc Natl Acad Sci USA 1990;87:8291-5.
59. Waters CA, Schimke PA, Snider CE et al. Interleukin 2 receptor-targeted cytotoxicity. Eur J Immunol 1990; 20:785-91.
60. Kiyokawa T, Shirono K, Hattori T et al. Cytotoxicity of interleukin 2-toxin toward lymphocytes from patients with adult T-cell leukemia. Cancer Res 1989; 49:4042-6.
61. Lorberboum-Galski H, Kozak RW, Waldmann TA et al. Interleukin 2 (IL-2) PE40 is cytotoxic to cells displaying either the p55 or p70 subunit of the IL-2 receptor. J Biol Chem 1988; 263:18650-6.
62. Gessain A, Barin F, Vernant JC et al. Antibodies to human T-lymphotropic virus type I in patients with tropical spastic paraparesis. Lancet 1985; ii:407-10.
63. Cruichshank E. A neuropathic syndrome of uncertain origin — review of 100 cases. West Indian Med J 1956; 39:592-5.
64. Mani K, Mani A, Montgomery R. A spastic paraplegia syndrome in South India. J Neurol Sci 1969; 9:179-99.

65. Montgomery R, Cruichshank E, Robertson W et al. Clinical and pathological observations on Jamaican neuropathy. Brain 1956; 87:425-62.
66. Osame M, Arima H, Narimatsu K et al. Epidemiostatistical studies of muscular atrophy in Southern Kyushu (Kagoshima and Okinawa prefectures). Jpn J Med 1975; 14:230-1.
67. Osame M, Usuku K, Izumo S et al. HTLV-I associated myelopathy, a new clinical entity. Lancet 1986; i:1031-2.
68. Osame M, Janessen R, Kubota H et al. Nationwide survey of HTLV-associated myelopathy in Japan: association with blood transfusion. Ann Neurol 1990; 28:51-6.
69. Roman GC, Osame M. Identity of HTLV-I associated tropical spastic paraparesis and HTLV-I-associated myelopathy [letter]. Lancet 1988; i:651.
70. Osame M. Review of WHO Kagoshima meeting and diagnostic guidelines for HAM/TSP. In: Balttner WA, ed. Human Retrovirology: HTLV. New York: Raven Press Ltd, 1990:191-7.
71. Johnson RT, Griffin DE, Arregui A et al. Spastic paraparesis and HTLV-I infection in Peru. Ann Neurol 1988; 23:S151-5.
72. Jacobson S, Gupta A, Mattson D et al. Immunological studies in tropical spastic paraparesis. Ann Neurol 1990; 27:129-56.
73. Itoyama Y, Minato S, Kira J et al. Altered subsets of peripheral blood lymphocytes in patients with HTLV-I associated myelopathy (HAM). Neurology 1988; 38:816-8.
74. Itoyama Y, Minato S, Kira J et al. Spontaneous proliferation of peripheral blood lymphocytes increased in patients with HTLV-I-associated myelopathy. Neurology 1988; 38:1302-7.
75. Jacobson S, Zaninovic V, Mora C et al. Immunological findings in neurological diseases associated with antibodies to HTLV-I: Activated lymphocytes in tropical spastic paraparesis. Ann Neurol 1988; 23:S196-200.
76. Yoshida M, Osame M, Kawai H et al. Increased replication of HTVL-I in HTLV-I-associated myelopathy. Ann Neurol 1989; 26:331-5.
77. Sonoda S, Yashiki S, Fujiyoshi T et al. Immunogenetic factors involved in the pathogenesis of adult T cell leukemia and HTLV-I-associated myelopathy. In: Takatsuki K, Hinuma Y, Yoshida M ed. Advances in Adult T Cell Leukemia and HTLV-I Research. Tokyo: Japan Scientific Societies Press, 1992:81-93.
78. Osame M, Igata A, Matsumoto M et al. HTLV-I-associated myelopathy (HAM), treatment trials, retrospective survey and clinical and laboratory findings. Hematol Rev 1990; 3:271-84.
79. Vernant JC, Maurs L, Gout O et al. HTLV-I-associated tropical spastic paraparesis in Martinique: A reappraisal. Ann Neurol 1988; 23(Suppl):S133-5.
80. Nishioka K, Maruyama I, Sato K et al. Chronic inflammatory

arthropathy associated with HTLV-1. Lancet 1989; i:441.
81. Sato K, Maruyama I, Maruyama Y et al. Arthritis in patients infected with human T lymphotropic virus type I. Clinical and immunopathologic features. Arthritis Rheum 1991; 34:714-21.
82. Nishioka K, Nakajima T, Hasunuma T et al. Rheumatic manifestation of human leukemic virus infection. Rheum Dis Clin North Am 1993; 19:489.
83. Kitajima I, Yamamoto K, Sato K et al. Deotection of HTLV-I proviral DNA and its gene expression in synovial cells in chronic inflammatory arthritis. J Clin Invest 1991; 88:1315-22.
84. Nakajima T, Aono H, Hasunuma T et al. Overgrowth of human synovial cells driven by the human T cell leukemia virus I tax gene. J Clin Invest 1993; 92:186-93.
85. Iwakura Y, Tosu M, Uoshida E et al. Induction of inflammatory arthropathy resembling rheumatoid arthritis in mice transgenic for HTLV-I. Science 1991; 253:1026.
86. Nussenblatt RB. Concepts of disease pathogenesis. In: Nussenblatt RB, Palestine AG, ed. Uveitis Fundamentals and Clinical Practice. Chicago: Yearbook Medical Publishers, 1989:21-53.
87. Mochizuki M, Watanabe T, Yamaguchi K et al. HTLV-I uveitis: a distinct clinical entity caused by HTLV-I. Jpn J Cancer Res 1992; 83:236-9.
88. Watanabe M, Muramatsu M, Tsuboi A et al. Differential response of NF-κB1 p105 and NF-κB2 p100 to HTLV-I encoded Tax. FEBS Lett 1994; 342:115-8.
89. Mochizuki M, Tajima K, Watanabe T et al. Human T lymphotropic virus type I uveitis. Br J Ophthalmol 1994; 78:149-54.
90. Morgan OS, Rodgers-Johnson P, Mora C et al. HTLV-I and polymyositis in Jamaica. Lancet 1989; ii:1184-7.
91. Kimura I, Tsubota T, Tada S et al. Presence of antibodies against adult T cell leukemia antigen in the patients with chronic respiratory diseases. Acta Medica Okayama 1986; 40:281-4.
92. Sugimoto M, Nakashima H, Watanabe S et al. T-lymphocyte alveolitis in HTLV-I-associated myelopathy. Lancet 1987; ii:1220.
93. Ohshima K, Kikuchi K, Masuda Y et al. HTLV-I-associated lymphadenopathy. Cancer 1992; 69:239-248.
94. LaGrenade L, Hanchard B, Fletcher V et al. Infective dermatitis of Jamaica children: a marker for HTLV-I infection. Lancet 1990; I:1345-7.

CHAPTER 3

HUMAN T CELL LEUKEMIA VIRUS TYPE I (HTLV-I)

Junji Yodoi

RETROVIRUS

HISTORICAL BACKGROUND

Retroviruses were first isolated from a chicken sarcoma in 1911 by Rous. In the 1950s, Gross isolated mammalian retroviruses from mouse leukemia.[1] Although a relationship between retroviruses and malignancies such as leukemia and lymphoma has been speculated in mice and other species, the leukemia virus has been a matter of debate for years.

Retroviruses may be differentiated by their ability to cause malignant and/or nonmalignant disorders or nonpathogenic disorders. A unique property of retroviruses is the potential of infectious agents to be transmitted in the germline as genetic Mendelian elements. This endogenous retrovirus is usually nonpathogenic.

Feline leukemia virus (FeLV) is known to cause both malignant and nonmalignant disease.[2] FeLV is the causative agent of both T-cell leukemia and AIDS-like T-cell immunodeficiency disease. Some retroviruses such as Gibbon ape leukemia virus (GaLV) are known to cause only malignant disease.[3]

Lentivirus is a class of retrovirus associated with nonmalignant diseases with chronic and progressive properties, including neurological diseases, hemolytic anemias and lung diseases. An example is sheep visna virus producing chronic neurological disease. Lentiviruses were previously known only in ungulates, however,

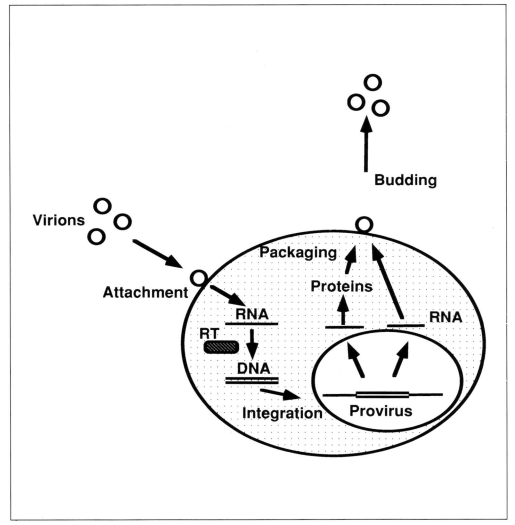

Fig. 3.1. Life cycle of retrovirus.

HIV associated with AIDS in humans and simian T-lymphotropic virus III (STLV-III) have been shown to be related to lentivirus.

BIOLOGY OF RETROVIRUSES

Retroviruses are enveloped single-strand RNA viruses. C-type retrovirus is a common retrovirus which reproduces by budding. When a retrovirus infects a cell, the reverse transcriptase (RT) catalyzes

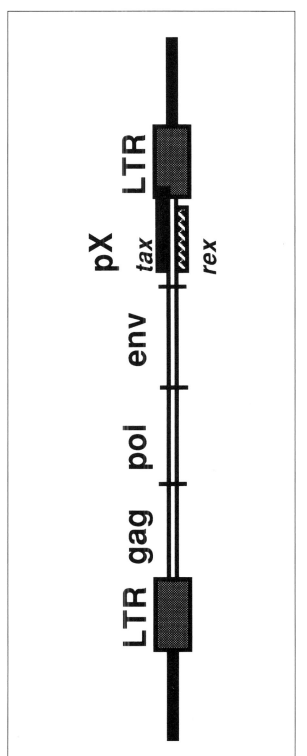

Fig. 3.2. Genetic structure of HTLV-I.

the transcription of the RNA genome into the DNA, which is known as the provirus.[4,5]

After the retrovirus integrates into the host cells, two viral elements called long terminal repeats (LTRs) define the integration sites of the virus, flanking the viral genome at the 3' and 5' ends. Expression of the provirus may vary depending on the host cell and environment. Viral RNA and proteins may be present in the cytoplasm and assembled at the cell membrane. Budding and release eventually occur, completing the cycle (Fig. 3.1).

HUMAN T LYMPHOTROPIC VIRUS TYPE I (HTLV-I)

As mentioned previously, HTLV-I isolated from American patients with CTCL was found to be the same as ATL from MT-1 cells, established by co-culture of ATL cells and cord lymphocytes.[6,7]

This retrovirus is a small RNA virus containing gag, pol and env genes without typical oncogenes. The gag gene codes for the internal structural proteins of the virus. The pol and env genes encode for reverse transcriptase and the outer glycoprotein envelope respectively (Fig. 3.2). In the early study of the structure of HTLV-I/ATL, Yoshida et al showed the presence of a peculiar open reading frame called pX.[8] The gene product of this pX, now called Tax, is a protein with various activities. Tax appears to interact not only with its own virus gene, but also with the host gene and its products.

Many laboratories were interested in the possible functional role of the Tax gene product on receptor expression because of the constitutive expression of IL-2R α chain in HTLV-I$^+$ T cell lines. Indeed, IL-2R α chain gene expression was apparently augmented by Tax, although the mechanism was initially not clear.[9-11] Sabe demonstrated that Tax apparently enhances the expression of IL-2R α chain, but not other components of the receptor complex, via transfection into YT cells.

Recently, Tax was shown to interact with host cell proteins including NF-κB.[12,13] The relationship between abnormal expression of the IL-2R α chain, ADF/TRX and Tax gene product in HTLV-I$^+$ cells is of particular interest. The emerging concept of redox regulation and oxidative stress participating in cell activation and cell death in relation to Tax is discussed in later chapters.

HUMAN T LYMPHOTROPIC VIRUS TYPE II (HTLV-II)

HTLV-II was first isolated from a cell line established from hairy cell leukemia.[14,15] There is no direct relationship between HTLV-II and ATL, despite a significant homology between HTLV-I and HTLV-II. Initial speculation that HTLV-II may be a causative agent of B cell malignancy has not been confirmed, partly due to difficulties in isolating this virus.

Recent epidemiological studies of human retroviruses have shown that HTLV-II positivity is high in South America. Despite initial reports on the possible association between HTLV-II and HAM/TSP, there remains no clear consensus about this.

REFERENCES

1. Gross L. Oncogenic Viruses. 3rd ed. Oxford: Pergamon Press 1983.
2. Krakower JM, Tonick SR, Gallagher RE et al. Antigenic characterization of a new Gibbon ape leukemia virus isolate: Seroepidemiologic assessment of an outbreak of Gibbon leukemia. Int J Cancer 1978; 22:715-20.
3. Jarret W., Martin B, Crighton W et al. Leukemia in the cat. Transmission experiments with leukemia (lymphosarcoma). Nature 1964; 202:566-7.
4. Temin HM and Mizutani S. RNA-dependent DNA polymerase in virions of Rous sarcoma virus. Nature 1970; 226:1211-1213.
5. Baltimore D. RNA-dependent DNA polymerase in virions of RNA tumour viruses. Nature 1970:226:1209-1211.
6. Poiesz BZ, Ruscetti FW, Gazdar AF et al. Detection and isolation of type C retrovirus particles from fresh and cultured lymphocytes of a patient with cutaneous T cell lymphoma. Proc Natl Acad Sci USA 1980; 77:7415-9.
7. Hinuma Y, Nagata K, Hanaoka M et al. Adult T-cell leukemia; Antigen in an ATL cell line and detection of antibodies to the antigen in human sera. Proc Natl Acad Sci USA 1981; 78:6476-80.
8. Seiki M, Hattori S, Hirayama Y et al. Human adult T-cell leukemia virus: Complete nucleotide sequence of the provirus genome integrated in leukemia cell DNA. Proc Natl Acad Sci 1983; 80:3618-22.
9. Inoue J, Seiki M, Taniguchi T et al. Induction of interleukin 2 receptor gene expression by p40x encoded by human T-cell leukemia virus type 1. EMBO J 1986; 5:2883-8.
10. Cross SL, Feinberg MB, Wolf JB et al. Regulation of the human interleukin-2 receptor α chain promotor: activation of a nonfunctional promotor by the transactivator gene of HTLV-I. Cell 1987; 49:47-56.

11. Sabe H, Tanaka A, Siomi H et al. Differential effects on expression of IL-2 receptors(p55 and p70) by the HTLV-I pX DNA. Int J Cancer 1988; 41:880-885.
12. Leung K, Nabel GJ. HTLV-1 transactivator induces interleukin-2 receptor expression through an NF-κB-like factor. Nature 1988; 333:776-778.
13. Cross S, Halden NF, Lenardo MJ et al. Functionally distinct NF-κB binding sites in the immunoglobulin κ and IL-2 receptor α chain genes. Science 1989; 244:466-468.
14. Saxon A, Stevens RH, Golde DW. T-lymphocyte variant of hairy cell leukemia. Ann Intern Med 1978; 88:323.
15. Gelmann EP, Franchini G, Manzari V et al. Molecular cloning of a new unique human T cell leukemia virus (HTLV-IIMo) In: Gallo RC, Essex M, Gross L eds. Human T-cell Leukemia/Lymphoma Virus. Cold Spring Harbor, NY. 1984:189-95.

CHAPTER 4

BIOLOGICAL PROPERTIES OF ATL CELLS

Takashi Uchiyama

CELL SURFACE PHENOTYPE AND FUNCTION IN VITRO

The T cell nature of leukemic cells was first demonstrated by their capability of forming rosettes with sheep red blood cells and their reactivity with rabbit anti-human T cell antisera.[1,2] Technological advances in the use of monoclonal antibodies to recognize various cell surface molecules, combined with flow cytometric analysis techniques, has made characterization of the cell surface phenotype of ATL cells easier and more precise. Analysis of cell surface phenotypes using various monoclonal antibodies has demonstrated that leukemic cells are $CD3^+CD4^+CD8^-$ in the majority (80-90%) of ATL cases and either $CD3^+CD4^+CD8^+$, $CD3^+CD4^-CD8^+$ or $CD3^+CD4^-CD8^-$ in the remaining 10-20% of ATL cases.[3-8] Several researchers have reported that the latter unusual phenotypes of leukemic cells were associated with a marked organomegaly, a bulky tumor mass, and a poor prognosis, compared with cases with the former typical cell surface phenotype.[8,9] Figure 4.1 shows the typical cell surface phenotype profile of fresh leukemic cells from ATL patients. This phenotype profile was obtained by flow cytometric analysis following immunofluorescence staining with a variety of monoclonal antibodies. They express CD3, CD4 and IL-2 receptor α and β chains, but not CD8.

Although expressed in a majority of normal peripheral T cells, CD7, an immunoglobulin superfamily molecule, is usually undetectable in peripheral blood leukemic cells.[10] Other cell surface antigens

Adult T Cell Leukemia and Related Diseases, edited by Takashi Uchiyama and Junji Yodoi. © 1995 R.G. Landes Company.

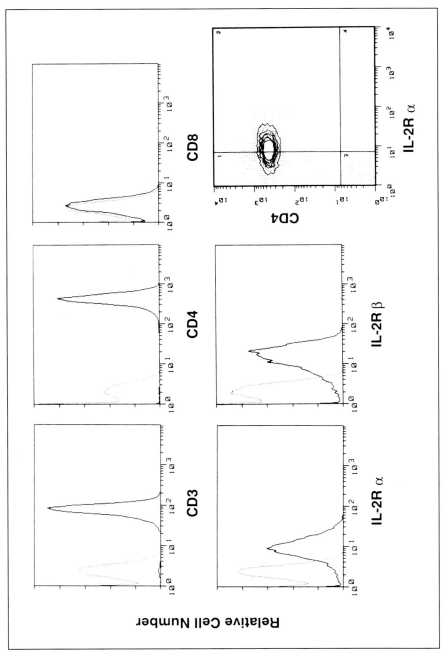

Fig. 4.1. Typical cell surface phenotype of leukemic cells from an ATL patient studied by flow cytometric analysis. The leukemic cells expressed CD3, CD4 and IL-2 receptor α and β chains, but not CD8.

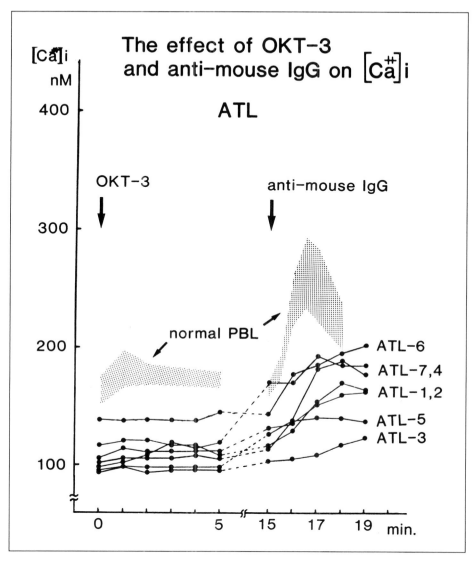

Fig. 4.2. The increase of intracellular free calcium concentration induced by stimulation with OKT3 antibody and anti-mouse IgG antibody is impaired in peripheral blood leukemic cells from ATL patients (ATL1-7) as compared with normal peripheral blood lymphocytes. Fluorescence intensity of Quin-2-loaded cells was measured with a fluorescence spectromonitor.

reported to be associated with cell activation or cell proliferation, such as Ki67, transferrin receptor (CD71), IL-2 receptor α chain and HLA-DR, are commonly expressed in leukemic cells. The expression of these cell activation- or proliferation-associated antigens has been reported to be higher in lymph node leukemic cells than in peripheral blood leukemic cells.[11]

T cell antigen receptors are the products of four rearranging genes (α, β, γ, δ). T cell receptors of immunocompetent T cells are noncovalently, but still intimately associated with CD3 complex which is composed of six peptide chains (CD3-γ-ε, δ-ε, ξ-ξ or ξ-η) and transduce the antigen recognition signal to the inside of the cell. In almost all cases ATL cells express T cell antigen receptor consisting of an α and β chain heterodimer and, in very rare cases the γδ type T cell receptor is expressed, together with CD3 molecules.[3-12] These results strongly suggest that leukemic cells are derived from peripheral mature T cells in most ATL cases. It has also been reported that CD3 molecules expressed on the ATL cell surface are fewer and their down-regulation more prominent in acute type than chronic type ATL.[13] This is well-correlated with IL-2R α chain expression. On the contrary, however, the amount of expressed mRNA of both T cell receptor α and β chain genes and CD3 γ, δ, ε, ζ, and η subunit genes is not decreased but rather increased. The decreased expression of CD3 molecules on the cell surface of ATL cells may be due to continuous or repetitive stimulation in vivo, although the precise nature of stimulation remains unclear.[11]

The activation of T cells initiated by ligand receptor interaction at the interface of T cells and antigen presenting cells is a complex process comprised of many cellular events. Stimulation of T cell receptor/CD3 complex induces activation of protein tyrosine kinase such as Fyn, Lck and Zap-70, which ultimately leads to activation of a variety of cellular genes, including the IL-2 gene.[14] On the other hand, stimulation of this complex induces rapid elevation in the concentration of intracellular free calcium, rapid breakdown and re-synthesis of inositol phospholipids, and activation of protein kinase C.[14] It would be interesting to see whether events induced by T cell activation are normal in leukemic cells from ATL patients, since data showing the dysfunction of HTLV-I-infected T cells has been reported. We measured a change in the intracellular free calcium concentration immediately following stimulation of CD3 complex by anti-CD3 antibodies. In contrast to the rapid increase in intracellular free calcium in normal peripheral blood T cells and leukemic T cells isolated from chronic lymphocytic leukemia patients, fresh leukemic cells examined from

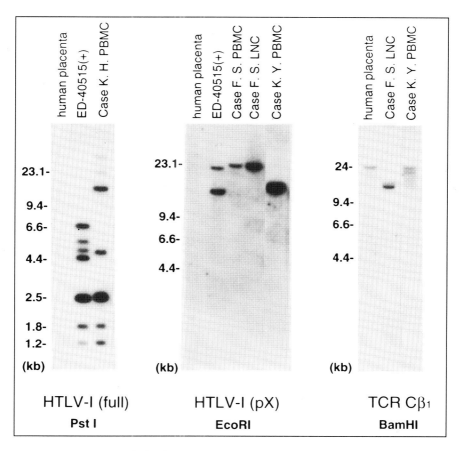

Fig. 4.3. Southern blot hybridization of leukemic cell DNA obtained from peripheral blood cells (PBMC) or lymph node cells (LNC) using full HTLV-I, pX of HTLV-I or Cβ1 of T cell receptor β chain gene. ED-40515+ cell line is a HTLV-I-infected cell line used as a positive control. Hybridization of cell DNA obtained from patients FS and KY showed a clear band(s) indicating that the cells examined are monoclonal.

10 ATL patients showed only a slight elevation (Fig. 4.2).[15] Furthermore, experiments with higher concentrations of anti-CD3 antibodies suggest that the attenuated increase in intracellular free calcium concentration was not primarily due to decreased expression of CD3 complex in ATL cells. The impaired early response of fresh ATL cells detected in these studies may be related to the immunodysfunction of HTLV-I-infected T cells.

Southern blot hybridization of leukemic cell DNA using a constant region of T cell receptor β chain gene (Cβ1) as a probe revealed distinct rearranged bands which indicated that leukemic cells from

ATL patients are a monoclonally expanded T cell population. These results confirm those from analysis of the HTLV-I provirus integration site in leukemic cell DNA (Fig. 4.3).[16] Leukemic T cells from almost all ATL patients show the rearrangement of T cell receptor α and γ chains associated with allelic deletion of the δ chain gene. It has not been reported, however, that particular variable segments of β chain gene are preferentially used by ATL cells. In other words, it does not appear that leukemic cells from ATL patients are derived from particular T cell clones.

In the extrathymic environment, the vast majority of T cells are naive resting cells which express high levels of CD45RA, B, C, a restricted group of high molecular weight isoforms of CD45 molecules. Once activated, T cells down-regulate CD45RA, B, C and express CD45RO, a low molecular weight isoform, as they differentiate to memory T cells.[17] ATL cells express CD45RO but not CD45RA in the majority of cases. These results suggest that HTLV-I may preferentially infect memory T cells or that HTLV-I-infected naive T cells may switch and express CD45RO upon activation or during cell transformation.

Fas/APO-1 is a 48 kD membrane protein and a member of the tumor necrosis factor (TNF) receptor/nerve growth factor (NGF) receptor family which mediates programmed cell death (apoptosis) following stimulation with its ligand or anti-Fas or anti-APO antibody.[18-21] It has been reported that fresh leukemic cells from ATL patients strongly expressed APO-1 when examined by flow cytometric analysis.[22] Incubation of ATL cells with anti-APO-1 antibody in vitro inhibited spontaneous and cytokine-mediated DNA synthesis. DNA isolated from incubated ATL cells exhibited polynucleosomal DNA fragmentation (DNA ladder) characteristic of apoptotic cell death. Given the high susceptibility of ATL cells to anti-APO-1 antibody and no curative chemotherapy available, Debatin et al suggested the use of anti-APO-1 antibody as a potential new strategy for the treatment of ATL.[22]

CYTOKINES AND OTHER BIOLOGICALLY ACTIVE PEPTIDES PRODUCED BY ATL CELLS

Interleukin-1 (IL-1) was discovered immunologically as a co-mitogen for thymocytes and lymphocytes and pathologically as an endogenous pyrogen produced by macrophages.[23,24] It is now known that IL-1 is produced by a variety of cells, including endothelial cells, renal mesangial cells, skin keratinocytes, glial cells and dendritic cells. The target cells of IL-1 include T cells, B cells, macrophages, fibroblasts, endothelial cells, osteoclasts, chondrocytes, hepatocytes, smooth muscle cells and keratinocytes. IL-1 is considered to be a major mediator of inflammation and is also involved in fever production, muscle proteolysis, bone resorption, wound healing and hematopoiesis. Human IL-1 exist in two functionally active forms: IL-1α and IL-1β, which share limited homology and differ in isoelectric points. A characteristic feature of IL-1 is the existence of its natural inhibitor, IL-1 receptor antagonist, which is produced by many different types of cells and exerts a potent IL-1-inhibitory effect.[25] Several reports showed that IL-1 α and β mRNA were produced by leukemic cells in a considerable proportion of ATL cases and also by the majority of HTLV-I-infected cell line cells.[26-28] IL-1 α and β produced by ATL cells may be responsible for neutrophilia sometimes observed in ATL cases and may also be involved in hypercalcemia.

Interleukin-2 (IL-2), discovered as a T cell growth-promoting factor,[29] is the principal cytokine which promotes the proliferation of T cells and natural killer (NK) cells. IL-2 also play an important role in the generation of cytotoxic killer T cells and lymphokine activated killer cells, the differentiation of NK cells and B cells, and the activation of macrophages to increase microbicidal activity.[30] The IL-2 produced by CD4+ T cells and in lesser quantities by CD8+ T cells functions as both an autocrine and paracrine growth factor for T cells. It has been reported that leukemic cells from some ATL patients produced IL-2 and their growth was inhibited by anti-IL-2R antibody in vitro, suggesting the role of IL-2 autocrine or paracrine mechanism in the neoplastic cell growth of ATL.[31,32] On the contrary, our studies demonstrated that peripheral blood leukemic cells proliferated in response to exogenous IL-2 only in 20-30% of the patients examined and IL-2 mRNA was not detected by Northern blot analysis in any of the 20 patients examined.[27,33] It is still controversial that IL-2 is constitutively produced by T cells upon infection with HTLV-I, and that the produced IL-2 promotes the growth of HTLV-I-infected T cells during the early phase of ATL development or is responsible for the neoplastic cell growth of ATL cells in vivo.

The cytokine interleukin-3 (IL-3) is produced by T cells, NK cells and mast cells, while GM-CSF is produced by T cells, NK cells, stromal cells, macrophages, endothelial cells and mast cells. Both cytokines promote growth and differentiation of hematopoietic progenitor cells and their receptors share a common β subunit.[34-36] In addition, the genes encoding IL-3 and GM-CSF map to within 10 kb of each other. Either gene could be activated by Tax of HTLV-I through interaction with NF-κB.[37] The expression of IL-3 mRNA in peripheral blood leukemic cells, however, was detectable in only 1 of 20 cases we examined.[27] In addition, both peripheral blood leukemic cells and lymph node neoplastic cells in this particular patient expressed HTLV-I viral RNA, which is very rare. It was not, however, apparent that the produced IL-3 was associated with or responsible for any clinical manifestations in this case. Generally, the production of IL-3 or GM-CSF by ATL cells is rarely detectable and their relation to pathogenesis or clinical features remains unknown.

Tumor necrosis factor (TNF) is a major inflammatory mediator with a broad spectrum of biological actions on many different target cells.[38,39] The major producers of TNF-α and TNF-β are monocytes/macrophages, but activated T cells also produce TNFs. There are two different types of TNF with considerable homology to each other: TNF-α and TNF-β (also called lymphotoxin). The genes of these two TNF types are closely linked and located within the major histocompatibility antigen complex. TNF-α promotes the proliferation of fibroblasts, enhances the synthesis of IL-1, IL-6, GM-CSF, M-CSF and interferon-gamma (IFN-γ), induces IL-2R α chain expression, and promotes B cell proliferation. TNF-α also elicits a wasting syndrome, enhances vascular permeability, induces pro-coagulant activity, promotes the adhesion of neutrophils, and activates osteoclasts. TNF-α and -β mRNA expression, TNF-β production by HTLV-I-infected cell lines, and elevated serum levels of TNF-β in hypercalcemic ATL patients have been reported.[40,41] The authors demonstrated the involvement of Tax of HTLV-I in the enhanced expression of TNF-β gene, probably through interaction with NF-κB, and suggested that the produced TNF-β might be involved in hypercalcemia.[41,42]

Mature TGF-β is a 25 kD homodimer produced by T cells, B cells, thymus, platelets, placenta and bone.[43] There are three

TGF-β isoforms in mammalian species. TGF-β causes bifunctional effects, both activation and suppression. TGF-β stimulates neovascularization, promotes the growth and activities of connective tissue cells, and plays a key role in wound healing. On the other hand, TGF-β is a potent immunosuppressive cytokine which suppresses both humoral and cell-mediated immunity. It suppresses IL-2-induced proliferation of T and B cells, and the generation of cytotoxic T cells and lymphokine-activated killer cells. It also suppresses the production of cytokines such as IL-1, IL-6 and TNF, and the expression of HLA-DR by antigen-presenting cells. It has been reported that TGF-β was produced by both HTLV-I-infected cell line cells and by leukemic cells from ATL patients,[44,45] probably through the effect of Tax. TGF-β may be one of the factors responsible for immunosuppression in ATL which often involves opportunistic infections by fungi, viruses and Pneumocystis carinii as serious complications. TNF and/or TGF-β produced by ATL cells in concert with PTHrP may mobilize calcium ion from the bone and result in bone resorption and severe hypercalcemia often encountered in ATL.

Regarding the immune function of ATL cells in vitro, we reported that peripheral blood leukemic cells from ATL patients showed a suppressive effect on pokeweed mitogen-induced differentiation of normal B cells and also that the culture supernatant of leukemic cells suppressed normal B cell differentiation.[46] It has subsequently been reported also by other groups that ATL cells or HTLV-I-infected T cells have suppressive activity.[6,47] It has not yet been determined, however, what mediates the suppressive activity of ATL cells. Cytokines such as TGF-β, which are produced by ATL cells and show immunosuppressive activity, may play a predominant role in the suppressive effects in vitro.

PTHrP is a protein comprised of 141 amino acids and shares sequence homology with PTH in the amino-terminal region. PTHrP shows effects on renal cell and osteoblast membranes similar to those of PTH, and it has also been suggested that its overproduction may induce hypercalcemia in some neoplasms.[48-51] Constitutive expression of PTHrP has been demonstrated in both HTLV-I-infected leukemic cells in ATL patients and HTLV-I-infected non-leukemic cells in HTLV-I carriers by a polymerase chain reaction technique.[52] Further studies revealed that Tax of HTLV-I caused transactivation of the PTHrP gene promoter.[52-54] Among the cytokines which have been suggested to produce hypercalcemia in ATL, the most important and predominant one appears to be PTHrP.

ATL-derived factor (ADF) was first defined as the IL-2R α chain(Tac)-inducing factor produced by HTLV-I-infected cell lines and subsequently proved to be a human homologue of the bacterial coenzyme thioredoxin, with thiol-dependent reducing activities.[55,56] ADF gene expression is enhanced in HTLV-I-infected T cell lines as well as Epstein-Barr virus-infected lymphoblastoid B cell lines. Its expression is also induced by a variety of stresses, including X-ray and ultraviolet irradiation, hydrogen peroxide, mitogens, and phorbol myristate acetate.[57] One of the important functions of ADF is the facilitation of intracellular protein-nucleotide interactions. It has been reported that recombinant ADF markedly enhances the binding of NF-κB to the target sequence in IL-2R α chain promoter as well as to the HIV-I LTR. One or more of ADF's functions may contribute to the pathogenesis of HTLV-I-associated diseases,[58] which is described in detail in a separate chapter.

Concerning the potential role of cytokines and/or neuromodulators in the pathogenesis of HTLV-I-related diseases of the nervous system, it has also been reported that Tax enhances the production of IL-1, IL-6, GM-CSF, TGF-β, and proenkephalin in gliomal cells.[59,60]

Acknowledgments

Our data summarized in this chapter are based on collaborative work with Drs. Toshio Hattori, Mitsuru Tsudo, Hiroshi Umadome, Shigeki Tamori and Taiichi Kodaka, to whom the author, Takashi Uchiyama, would like to express his sincere thanks.

References

1. Yodoi J, Takatsuki K, Masuda T. Two cases of T-cell chronic lymphocytic leukemia in Japan. N Engl J Med 1974; 290:572-3.
2. Uchiyama T, Yodoi J, Sagawa K et al. Adult T-cell leukemia: clinical and hematologic features of 16 cases. Blood 1977; 50:481-92.
3. Hattori T, Uchiyama T, Toibana T et al. Surface phenotype of Japanese ATL cells characterized by MoAbs. Blood 1981; 58:645-7.
4. Uchiyama T, Hattori T, Wano Y et al. Cell surface phenotype and in vitro function of adult T cell leukemia cells. Diag Immunol

1983; 1:150-4.
5. Schnitzer B, Lovett EJ 3rd, Hudson JL et al. Adult T-cell leukemia-lymphoma with unusual phenotype [letter]. Lancet 1982; ii:1273-4.
6. Yamada Y. Phenotypic and functional analysis of leukemic cells from 16 patients with adult T-cell leukemia/lymphoma. Blood 1983; 61:192-9.
7. Tamura K, Unoki T, Sagawa K et al. Clinical features of OKT4+/OKT8+ adult T cell leukemia. Leuk Res 1985; 9:1353-9.
8. Kamihira S, Sohda H, Atogami S et al. Phenotypic diversity and prognosis of adult T cell leukemia. Leuk Res 1992; 16:435-41.
9. Kamihira S. Hemato-cytological aspects of adult T-cell leukemia. In: Takatsuki K, Hinuma Y, Yoshida M ed. Advances in Adult T Cell Leukemia and HTLV-I Research. Tokyo: Japan Scientific Societies Press, 1992:17-32.
10. Haynes BF, Denning SM, Singer KH et al. Ontogeny of T cell precursors: a model for the initial stages of human T cell development. Immunol Today 1989; 10:87-91.
11. Shirono K, Hattori T, Hata H et al. Profiles of expression of activated cell antigens on peripheral blood and lymph node cells from different clinical stages of adult T cell leukemia. Blood 1989; 73:1664-71.
12. Hattori T, Asou N, Suzushima H et al. Leukemia of novel gastrointestinal T-lymphocyte population infected with HTLV-I. Lancet 1991; 337:76-7.
13. Matsuoka M, Hattori T, Chosa T et al. T3 surface molecules on adult T cell leukemia cells are modulated in vivo. Blood 1986; 67:1070-6.
14. Weiss A. T lymphocyte activation. In: Paul WE ed. Fundamental Immunology. 3rd ed. New York: Raven Press, 1993:467-504.
15. Tamori S, Uchiyama T, Umadome H et al. Increase in cytoplasmic free calcium concentration initiated by T3 antigen stimulation is impaired in adult T cell leukemia cells. Leuk Res 1988; 12:357-63.
16. Maeda M, Shimizu A, Ikuta K et al. Origin of human T-lymphotrophic virus I-positive T cell lines in adult T cell leukemia. J Exp Med 1985; 162:2169-74.
17. Thomas ML. The leukocyte common antigen family. Ann Rev Immunol 1989; 7:339-69.
18. Yonehara S, Ishi A, Yonehara M et al. A cell killing monoclonal antibody (anti-Fas) to a cell surface antigen is co-downregulated with the receptor of tumor necrosis factor. J Exp Med 1989; 169:1747-56.
19. Trauth BC, Klas C, Peters AMJ et al. Monoclonal antibody-mediated tumor regression by induction of apoptosis. Science 1989; 245:301-5.

20. Itoh N, Yonehara S, Ishii A et al. The polypeptide encoded by the cDNA for human cell surface antigen Fas can mediate apoptosis. Cell 1991; 66:233-43.
21. Suda T, Takahashi T, Goldstein P et al. Molecular cloning and expression of the Fas ligand, a novel member of the tumor necrosis factor. Cell 1993; 75:1169-78.
22. Debatin KM, Goldmann CK, Waldmann TA et al. APO-1-induced apoptosis of leukemia cells from patients with adult T cell leukemia. Blood 1993; 81:2972-7.
23. Oppenheim JJ, Kovacs EJ, Matsushima K et al. There is more than one interleukin 1. Immunol Today 1986; 7:45-56.
24. Dinarello CA. Interleukin 1 and its biologically related cytokines. Adv Immunol 1989; 44:153-205.
25. Eisenberg SP, Evans RJ, Arend WP et al. Primary structure and functional expression from complementary DNA of a human interleukin 1 receptor antagonist. Nature 1990; 343:341-6.
26. Wano Y, Hattori T, Matsuoka M et al. Interleukin 1 gene expression in adult T cell leukemia. J Clin Invest 1978; 80:911-6.
27. Kodaka T, Uchiyama T, Umadome H et al. Expression of cytokine mRNA in leukemic cells from adult T cell leukemia patients. Jpn J Cancer Res 1989; 80:531-6.
28. Noma T, Nakakubo H, Sugita M et al. Expression of different combinations of interleukins by human T cell leukemic cell lines that are clonally related. J Exp Med 1989; 169:1853-8.
29. Morgan DA, Ruscetti FW, Gallo RC. Selective in vitro growth of T-lymphocytes from normal human bone marrows. Science 1976; 193:1007-8.
30. Smith KA. Interleukin-2. Inception, impact, and implication. Science 1988;240:1169-76.
31. Arima N, Daitoku Y, Ohgaki S et al. Autocrine growth of interleukin 2 producing leukemic cells in a patient with adult T cell leukemia. Blood 1986; 68:776-82.
32. Goebels N, Waawe I, Pifzenmaier K et al. IL-2 production in human T lymphotropic virus I-infected leukemic T lymphocytes analyzed by in situ hybridization. J Immunol 1998; 141:1231-5.
33. Uchiyama T, Hori T, Tsudo M et al. Interleukin-2 receptor (Tac antigen) expressed on adult T cell leukemia cells. J Clin Invest 1985; 76:446-53.
34. Ihle JN and Weinstein Y. Immunological regulation of hematopoietic/lymphoid stem cell differentiation by interleukin 3. Adv Immunol 1986; 39:1-50.
35. Shrader JW, Lewis SL, Clark-Lewis I et al. The persistent(P) cell: histamine content, regulation by a T cell derived factor, origin from a bone marrow precursor, and relationship to mast cells. Proc Natl Acad Sci USA 1981; 78:323-7.
36. Nicola NA. Hematopoietic cell growth factors and their receptors.

Ann Rev Biochem 1989; 58:45-77.
37. Miyatake S, Seiki M, Malefijt RD et al. Activation of T cell-derived lymphokine genes in T cells and fibroblasts: effects of human T cell leukemia virus type I p40^x protein and bovine papilloma virus encoded E2 protein. Nucleio Acids Res 1988; 16:6547-66.
38. Old LJ. Tumor necrosis factor. In: Bonavida B, Granger G ed. Tumor Necrosis Factor: Structure, Mechanism of Action, Role in the Disease and Therapy. Basel: Karger, 1990:1-30.
39. Ruddle NH, Millet I, Picarella DD et al. TNF-α(cachectin) and TNF-β(lymphotoxin): 25 years of progress. In: Fiers W, Buurman WA ed. Tumor Necrosis Factor: Molecular and Cellular Biology and Clinical Relevance. Basel: Karger, 1993:1-9.
40. Salahuddin SZ, Markham PD, Linder SG et al. Lymphokine production by cultured human T cells transformed by human T-cell leukemia-lymphoma virus-I. Science 1984; 223:703-7.
41. Tschachler E, Robert-Guroff M, Gallo RC et al. Human T-lymphotropic virus I-infected T cells constitutively express lymphotoxin in vitro. Blood 1989; 73:194-201.
42. Tschachler E, Bohnlein E, Felzmann S et al. Human T-lymphotropic Virus Type I tax Regulates the Expression of the Human Lymphotoxin Gene. Blood 1993; 81:95-100.
43. Sporn MB, Roberts AB. Transforming growth factor-β: multiple actions and potential clinical applications. JAMA 1989; 262:938-41
44. Niitsu Y, Urushizaki Y, Koshida Y et al. Expression of TGF-β gene in adult T cell leukemia. Blood 1988; 71:263-6.
45. Kim S-J, Kehrl JH, Burton J et al. Transactivation of the transforming growth factor β1(TGF-β1) gene by human T lymphotropic virus type 1 tax: a potential mechanism for the increased production of TGF-β1 in adult T cell leukemia. J Exp Med 1990; 172:121-9.
46. Uchiyama T, Sagawa K, Takatsuki K et al. Effect of adult T-cell leukemia cells on pokeweed mitogen-induced normal B-cell differentiation. Clin Immunol Immunopathol 1978; 10:24-34.
47. Morimoto C, Matsuyama T, Oshige C et al. Functional and phenotypic studies of Japanese adult T cell leukemia cells. J Clin Invest 1985; 75:836-43.
48. Stewart AF, Horst R, Deftos LJ et al. Biochemical evaluation of patients with cancer-associated hypercalcemia: evidence for humoral and nonhumoral groups. N Engl J Med 1980; 303:1377-83.
49. Suva LJ, Winslow GA, Wettenhall REH et al. A parathyroid hormone-related protein implicated in malignant hypercalcemia: Cloning and expression. Science 1987; 237:893-6.
50. Mangin M, Webb AC, Dreyer B et al. Identification of a cDNA encoding a parathyroid hormone-like peptide from a human tumor associated with humoral hypercalcemia of malignancy. Proc Natl Acad Sci USA 1988; 85:597-601.

51. Broadus AE, Mangin M, Ikeda K et al. Humoral hypercalcemia of cancer. Identification of a novel parathyroid hormone-like peptide. N Engl J Med 1988; 319:556-63.
52. Watanabe T, Yamaguhci K, Takatsuki K et al. Constitutive expression of parathyroid hormone-related protein gene in human T cell leukemia virus type I(HTLV-I) carriers and adult T cell leukemia patients that can be trans-activated by HTLV-I tax gene. J Exp Med 1990; 172:759-65.
53. Ikeda K, Okazaki R, Inoue D et al. Interleukin-2 increases production and secretion of parathyroid hormone-related peptide by human T cell leukemia virus type I-infected T cells: possible role in hypercalcemia associated with adult T cell leukemia. Endocrinology 1993; 132:2551-6.
54. Ejima E, Rosenblatt JD, Massari M et al. Cell-type-specific transactivation of the parathyroid hormone-related protein gene promoter by the human T-cell leukemia virus type I (HTLV-I) tax and HTLV-II tax proteins. Blood 1993; 81:1017-24.
55. Teshigawara K, Maeda M, Nishino K et al. Adult T leukemia cells produce a lymphokine that augments interleukin 2 receptor expression. J Mol Cell Immunol 1985; 2:17-26.
56. Tagaya Y, Maeda Y, Mitsui A et-al. ATL-derived factor (ADF), an IL-2 receptor/Tac inducer homologous to thioredoxin; possible involvement of dithiol-reduction in the IL-2 receptor induction. EMBO J 1989; 8:757-64.
57. Yodoi J, Tursz T. ADF: A endogenous reducing protein homologous to thioredoxin: Involvement in lymphocyte immortalization by HTLV-I and EBV. Adv Cancer 1991; 57:382-410.
58. Yodoi J, Uchiyama T. Diseases associated with HTLV-I: virus, IL-2 receptor dysregulation and redox regulation. Immunol Today 1992; 13:405-11.
59. Sawada M, Suzumura A, Kondo N et al. Induction of cytokines in glial cells by transactivator of human T-cell lymphotropic virus type I. FEBS-Lett 1992; 313:47-50.
60. Joshi JB, Dave HP. Transactivation of the proencephalon gene promoter by the Tax1 protein of human T-cell lymphotropic virus type I. Pro Natl Acad Sci USA 1992; 89:1006-10.

= CHAPTER 5 =

IL-2/IL-2 RECEPTOR SYSTEM IN ATL

Takashi Uchiyama

INTERLEUKIN-2

Interleukin-2 (IL-2) was first discovered as a T cell growth factor contained in culture medium of PHA-stimulated peripheral blood leukocytes which selectively promoted the growth of T cells in the culture of human bone marrow cells.[1,2] IL-2 is a markedly hydrophobic glycoprotein composed of 133 amino acids with an approximate molecular weight of 15,000-17,000.[3-5] It is produced from T cells activated by various activators including mitogens, antigens presented by antigen presenting cells, calcium ionophore + PMA, antibodies to CD2 and crosslinking antibodies of T cell receptor and CD3 complex. A limited variety of different kinds of cells are affected by IL-2.[4,6] In T cells, IL-2 promotes progression through G1 phase of cell cycle, promotes proliferation, induces c-myc and c-myb expression, stimulates cytolytic activity, and stimulates the production of various cytokines including interferon-gamma (IFN-γ). IL-2 promotes proliferation of large granular lymphocytes and enhances natural killer (NK) cell activity and LAK activity.[7] It also stimulates the proliferation of B cells[8] and enhances the microbicidal activity of monocytes.[9]

The stimulation of T cell receptor/CD3 complex induces a series of intracellular events which are involved in or associated with the transduction of a T cell activation signal (Fig. 5.1). Two major early events in signal transduction pathways triggered by T cell

Adult T Cell Leukemia and Related Diseases, edited by Takashi Uchiyama and Junji Yodoi. © 1995 R.G. Landes Company.

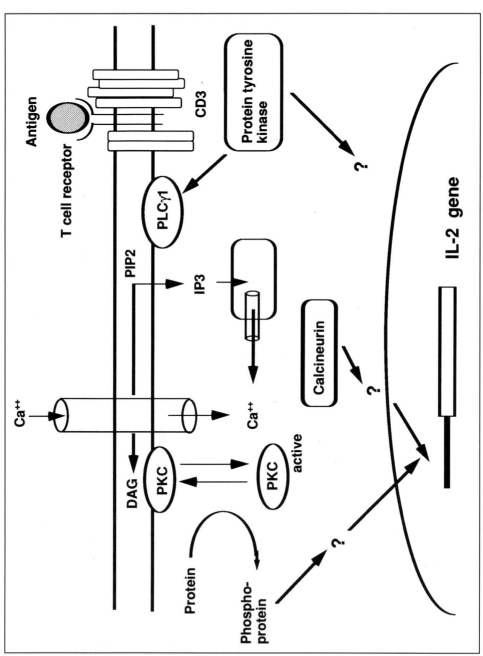

Fig. 5.1. Intracellular signaling pathway of T cells following T cell receptor-mediated stimulation.

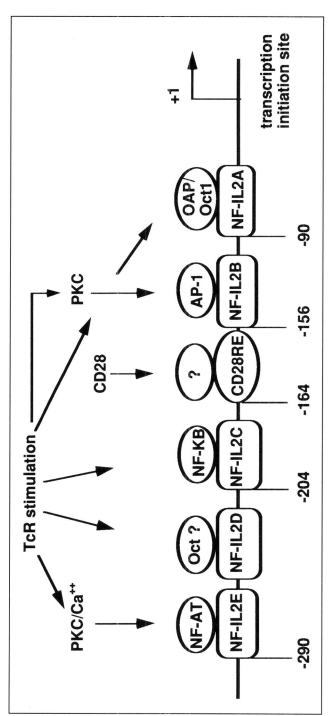

Fig. 5.2. 5' upstream regulatory regions of IL-2 gene.[12-16]

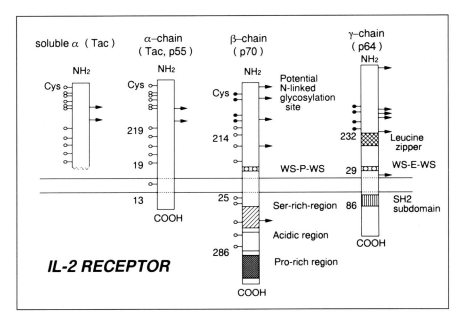

Fig. 5.3. Schematic structure of three subunits of IL-2 receptor.[25,26,29,32]

receptor/CD3 complex stimulation are: (1) activation of T cell receptor/CD3 complex-associated phosphotyrosine kinases including Fyn, Lck and ZAP-70; and (2) a rapid increase in intracellular free calcium concentration, followed by the activation of protein kinase C.[10,11] Both the increase of intracellular free calcium and the activation of protein kinase C are required for transcriptional activation of IL-2 gene in resting T cells. It has been reported that the 5' upstream regulatory region of the IL-2 gene contains five T cell receptor-responsive (NF-IL2A, NF-IL2B, NF-IL2C, NF-IL2D and NF-IL2E) elements and one CD28-responsive element (CD28RE).[12-16] The NF-IL2A, NF-IL2B, NF-IL2C and NF-IL2E sites bind Oct-1, AP-1 or related proteins, NF-κB and AP-3, and NF-AT transcription factor, respectively (Fig. 5.2). NF-AT complex contains c-Fos or c-Fos/Jun or related proteins in addition to NF-AT component, the cytoplasmic subunit of which has recently been cloned in humans.[17] NF-AT complex, in response to calcium signals, translocates from cytoplasm into the nucleus following dephosphorylation by the calcium/calmodulin-dependent protein phosphatase calcineurin, which is the target of immunosuppressants, cyclosporin A and FK506.[18]

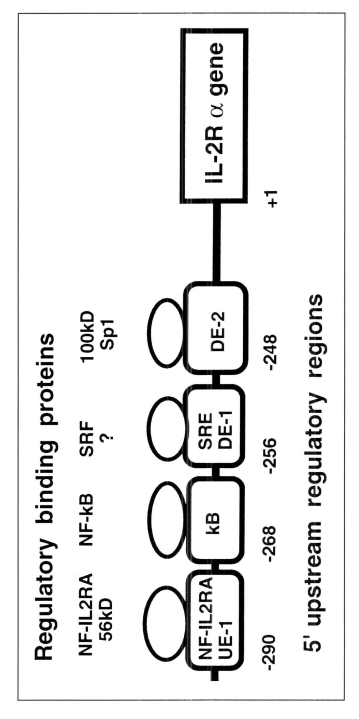

Fig. 5.4. 5' upstream regulatory regions of IL-2 receptor α chain gene and regulatory binding proteins.[34-36]

Stimulation of CD28 in the presence of a rapid increase in intracellular free calcium and activation of protein kinase C also induces a nuclear factor which can bind to CD28RE of the regulatory region of IL-2 gene and activate the transcription.[19]

Transcriptional regulation described above is certainly a major mechanism by which IL-2 gene expression is controlled. However, post-transcriptional regulation may also be involved in the regulation of IL-2 gene expression as reported in other cytokines. Transcripts of most cytokine genes contain an AU-rich 3' untranslated portion that confers instability to the mRNA. It has been reported that stimulation of T cells with phorbol esters or CD28 stabilizes cytokine mRNA presumably through an AU-rich untranslated portion.[20]

INTERLEUKIN-2 RECEPTOR (IL-2R)

IL-2 receptor (IL-2R) was first detected by specific binding of radiolabeled IL-2 to the cell surface.[21] Its cDNA was cloned subsequent to characterization using anti-Tac monoclonal antibody which recognizes IL-2R.[22-26] The molecule recognized by anti-Tac antibody is now known as α chain, which is one of three subunits of IL-2R. IL-2R consists of three different subunits, namely α, β and γ chains (Fig. 5.3).[23-32] The α chain is a glycoprotein with a protein backbone of 251 amino acids with an approximate molecular weight of 55,000. The extracytoplasmic portion composed of 219 amino acids and attached carbohydrates contains an IL-2 binding portion. The hydrophobic portion of the α chain is a membrane spanning portion and intracytoplasmic portion which consists of only 13 amino acids and is too short to mediate signal transduction. The α chain appears to facilitate the formation of the complex composed of IL-2, α chain, β chain and γ chain.[33]

The expression of IL-2R α chain is predominantly transcriptionally regulated. The regulatory region of IL-2R α chain gene (positions -290 to -240, relative to the major transcription start site) contains at least four elements to which sequence-specific nuclear factors bind (Fig. 5.4).[34-36] NF-κB, which binds to the κB site of the promoter region, has been reported to play a key role in the activation of IL-2R α chain gene by Tax of HTLV-I and appears to play a critical role in the regulation of the gene in normal T cell activation. In addition to NF-κB, nuclear factors binding to either serum response element (SRE)-like site or NF-IL2RA may be differentially important in IL-2R α chain gene transcription.

Both β chain and γ chain are members of cytokine receptor superfamily which is characterized by four conserved χψστεινες and WSXWS motif. The β chain consists of 525 amino acids and has an approximate molecular weight of 70,000.[29] The cytoplasmic region is composed of 286 amino acids and contains a serine-rich region and an acidic region. These regions appear to play a critical role in signal transduction, as demonstrated by expression studies with deletion mutant cDNA.[37,38] To our surprise, the receptor for IL-15, for which cDNA has recently been cloned and partially characterized, appears to employ IL-2R β and γ chains as subunits.[39]

The γ chain, another subunit of IL-2R, is also a membrane type receptor glycoprotein composed of 347 amino acids with an approximate molecular weight of 64,000.[32] IL-2R γ chain has also been demonstrated to be essential for IL-2 signal transduction by similar expression studies using deletion mutant cDNA. It has recently been revealed by expression studies with chimeric receptor cDNAs that heterodimerization of the β- and γ-chain cytoplasmic domains is required for IL-2 signaling.[40] It is interesting that the γ chain is shared among receptors for IL-2, IL-4, IL-7, IL-9, IL-13 and IL-15.[41,42] The β chain is constitutively expressed in CD8+ T cells and large granular lymphocytes and is also transiently induced to be expressed on activation of T cells.[43-45] On the other hand, γ chain is constitutively expressed in lymphoid cell lines and unstimulated peripheral blood lymphocytes.[32]

There are four different IL-2Rs in terms of the affinity to the ligand.[32,46,47] The high affinity, the pseudohigh affinity, the intermediate affinity and the low affinity IL-2R are considered to be composed of αβγ, αβ, βγ and α chains, respectively (Fig. 5.5). Binding kinetics studies demonstrated that the association and dissociation of IL-2 were rapid in α chain and slow in β chain,[46,47] and that γ chain was responsible for the slow dissociation of IL-2 bound to β chain.[33]

In addition to the membrane form of IL-2R, it was found that the soluble form of the α chain (s-IL-2R α), which is devoid of membrane spanning and cytoplasmic region, is released from activated T cells and HTLV-I-infected T cells.[48] It can be detected by a sandwich ELISA using two anti-IL-2R α chain monoclonal antibodies which recognize different epitopes of the extracytoplasmic region. The mRNA encoding the soluble form of α chain has not been detected so far. Although the relevant enzyme has not yet

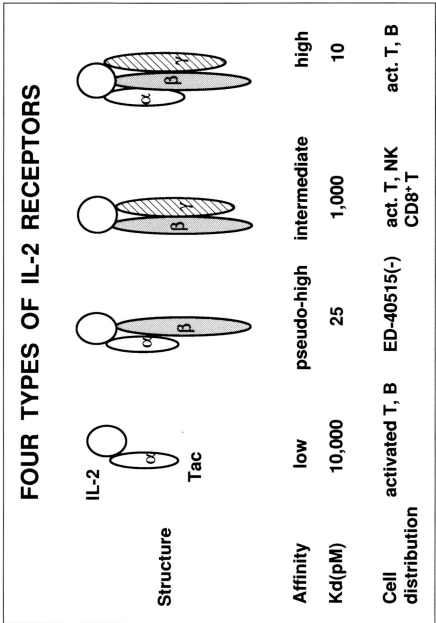

Fig. 5.5. Four different types of IL-2 receptor.

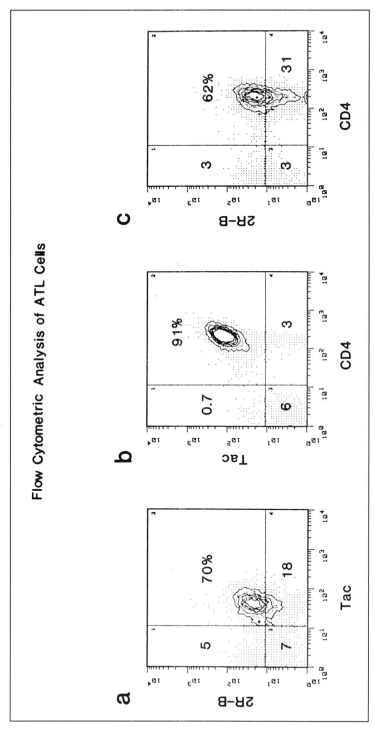

Fig. 5.6. Expression of IL-2 receptor α and β chain on peripheral blood leukemic cells from an ATL patient demonstrated by flow cytometric analysis.

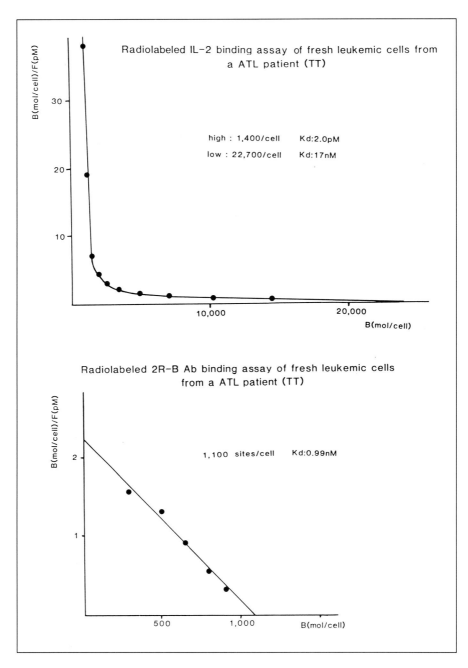

Fig. 5.7. Radiolabeled IL-2 binding assay of fresh leukemic cells from an ATL patient (upper panel). Scatchard plot analysis showed 1,400/cell high affinity (Kd: 2.0 pM) and 22,700/cell low affinity (Kd: 17 nM) receptors. Radiolabeled anti-IL-2R β chain antibody binding assay (lower panel) revealed 1,100/cell β chains.

been identified, it has been suggested that the soluble form of α chain is produced by the proteolysis of the membrane form α chain. It is known that the sera of healthy individuals contains a small amount of soluble α chain, however, its physiological role remains unclear.

IL-2/IL-2R EXPRESSION IN ATL

IL-2R α and β chain expression on the cell surface of leukemic cells was examined by immunofluorescence staining using monoclonal antibodies such as anti-Tac antibody and 2R-β antibody followed by flow cytometric analysis.[49-52] In contrast to the lack of expression on unstimulated CD4+ peripheral blood T cells from normal individuals, peripheral blood leukemic cells from almost all patients with ATL were found to bear IL-2R α and β chain (Fig. 5.6). The HTLV-I-infected cell lines usually expressed a larger number of IL-2Rs judging from fluorescence intensity. The number of IL-2Rs expressed on the cell surface is estimated by Scatchard plot analysis following radiolabeled IL-2 binding assay.[51] Leukemic cells isolated from 14 ATL patients expressed 150-1,500/cell high affinity (Kd:10-25 pM) and 1,600-22,700/cell low affinity (Kd:17-77 nM) receptors (Fig. 5.7). HTLV-I-infected cell lines such as Hut102 cells express a much higher number of high and low affinity IL-2 receptors. Similar data on IL-2R α and β chain expression in leukemic cells from ATL patients and HTLV-I-infected cell lines were also reported from other investigators.[53-57] To date, neither the obvious amplification nor gross rearrangement of IL-2R α and β chain genes has been reported, although the possibility of a minute change such as a point mutation of the gene has not been excluded.

Some researchers have reported data which strongly suggest that leukemic cells from some ATL patients proliferated by IL-2 autocrine mechanism.[58] However, IL-2 production by peripheral blood leukemic cells was hardly detectable in 20 cases we studied by Northern blot hybridization and any HTLV-I-infected cell lines examined did not produce a detectable amount of IL-2.[59] In addition, the spontaneous uptake of ^{3}H-thymidine by freshly isolated peripheral blood leukemic cells was usually very low. These results suggest that the IL-2 autocrine mechanism does not support neoplastic cell growth of at least peripheral blood leukemic cells from ATL patients. In vitro transfection experiments disclosed that tax

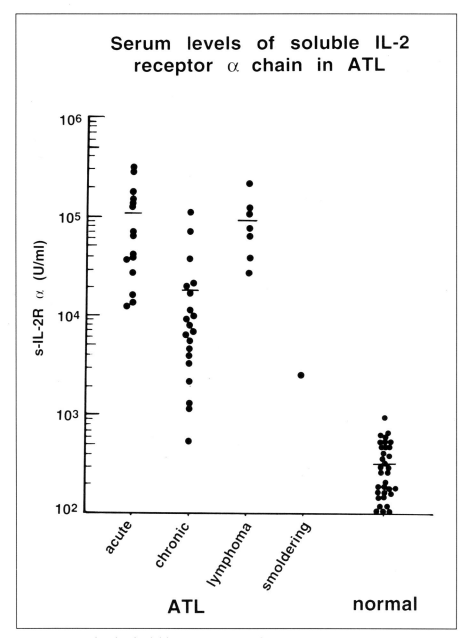

Fig. 5.8. Serum levels of soluble IL-2 receptor α chain (s-IL-2R α) in ATL.

gene of HTLV-I introduced into cells was capable of inducing the expression of IL-2, although less strongly than that of IL-2R α chain. It was reported that the product of tax gene (p40^{tax}, Tax) enhances the transcription of IL-2 gene through interaction with a transcription factor, NF-κB.[60]

The expression of IL-2R α chain increases as leukemic cells are cultured in vitro. Comparative studies on the expression of HTLV-I, IL-2R α chain and other cellular genes in short-term cultured leukemic cells from ATL patients revealed a close association between HTLV-I viral RNA expression and IL-2R α chain mRNA.[61] These findings strongly suggest that HTLV-I infection induces IL-2R α chain expression. With respect to the oncogenic role of HTLV-I, the gene in which people studying ATL and HTLV-I have been most interested is probably the tax gene of HTLV-I.[62] Subsequent to the demonstration of a transacting transcriptional activity of pX products(p40^{tax}, Tax),[63,64] the induction of the IL-2R α chain gene expression by Tax has been established by transfection experiments.[60,65-67] The transacting transcription-enhancing activity of Tax is not due to direct interaction of Tax with specific DNA sequences of IL-2R α chain gene, but is mediated by a cellular transcription factor, NF-κB which binds to the 12 base pair motif of the promoter region of the α chain following interaction with Tax. In addition to the direct transcription enhancing activity of Tax/NF-κB complex, recent studies on Tax have shown that Tax also facilitates the shift of NF-κB from cytoplasm to nucleus by binding to and dissociating I-κB from cytoplasmic NF-κB/I-κB complex resulting in the enhancement of IL-2R α chain gene transcription.[68]

As mentioned above, IL-2R β chain is expressed not only on the cell surface of freshly isolated leukemic T cells, but also on HTLV-I-infected cell line cells. This is in contrast to lack of IL-2R β expression on unstimulated CD4$^+$ T cells, which are normal counterparts of ATL cells.[52,56,57] It remains controversial, however, whether HTLV-I enhances expression of IL-2R β chain gene expression. The possible association of IL-2R γ chain expression with HTLV-I infection requires further study.

It has been reported that serum levels of s-IL-2R α are elevated in various diseases, including hematopoietic neoplasms, autoimmune diseases, graft rejection, infections, and burns.[69] Among diseases

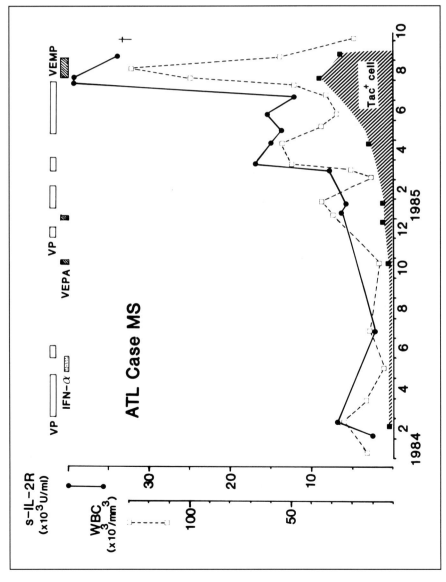

Fig. 5.9. Serum levels of soluble IL-2 receptor α chain (s-IL-2R α) measured throughout the clinical course of ATL case MS.

characterized by elevated s-IL-2R α, the most remarkable is ATL.[70-72] Concentrations of more than 385 ng/ml (100,000 U/l), as compared to normal levels (0.5-2.3 ng/ml), are frequently observed in acute and lymphoma type ATL, while mild to moderate elevations occur in chronic and smoldering type ATL. Characteristics such as markedly elevated levels of serum s-IL-2R α and expression of IL-2R α chain on CD4⁺ T cells, in conjunction with other clinical manifestations, are useful in the diagnosis of ATL (Fig. 5.8). Furthermore, s-IL-2R α level is a useful indicator in follow-up of patients throughout the clinical course and in the evaluation of therapy (Fig. 5.9). In one case (case MS in Fig. 5.9), changes in serum levels of s-IL-2R α were roughly parallel to those of LDH and peripheral blood T cells expressing both CD4 and IL-2R α chain, which are probably the leukemic cells. In another case, a lymphoma type ATL ran a rapid downhill course following diagnosis and had severe hypercalcemia with marked osteolytic lesions. The measurement of serum s-IL-2R α provides useful information on the total mass or number of neoplastic cells in the entire body, especially in cases where leukemic cells are scarcely detectable in peripheral blood. It has been reported that s-IL-2R α chain-coupled supports have the ability to bind IL-2 in vitro, and s-IL-2R α chain in solution may also be able to bind IL-2, although the affinity is much lower than the low affinity IL-2R expressed on the cell surface.[73] A large amount of s-IL-2R α chain present in the sera of ATL patients potentially inhibit IL-2 function by binding to IL-2. This may be one of the mechanisms underlying the immune deficiency state in ATL.

An intriguing hypothesis for the mechanism of the leukemogenesis of ATL is that constitutively expressed IL-2 and its receptor, induced by HTLV-I infection, are critically involved in either the continuous or repetitive cell cycle progression of cell cycle of HTLV-I-infected and Tax producing cells. During a long period when HTLV-I-infected cells undergo multiple and repetitive cell proliferation, cellular events which seriously affect cell growth and/or cell death may accumulate and eventually lead to uncontrollable cell growth. It should be noted, however, that Tax affects the expression of a variety of cellular genes other than IL-2 and IL-2Rα, the products of which are closely related to cell growth or cell death. Such products include GM-CSF, TNF-β, TGF-β, products of c-fos, c-jun and egr, and β-polymerase.[74-79] Abnormal expression of these molecules is also potentially involved in the leukemogenesis of ATL.

ACKNOWLEDGMENTS

Our data described in this chapter were obtained from collaborative studies with Drs. Mitsuru Tsudo, Yuji Wano, Hiroshi Umadome, Toshiyuki Hori, Masanori Kamio, Taiichi Kodaka, Nobuyoshi Arima and Michiyuki Maeda to whom the author, Takashi Uchiyama, would like to express sincere thanks.

REFERENCES

1. Morgan DA, Ruscetti FW, Gallo RC. Selective in vitro growth of T-lymphocytes from normal human bone marrows. Science 1976; 193:1007-8.
2. Ruscetti FW, Morgan DA, Gallo RC. Functional and morphologic characterization of human T cells continuously grown in vitro. J Immunol 1977; 119:131-8.
3. Gillis S, Ferm MM, Ward OU et al. T-cell growth factor: parameters of production and a quantitative microassay for activity. J Immunol 1978; 120:2027-31.
4. Smith KA. T-cell growth factor. Immunol Rev 1980; 51:337-57.
5. Taniguchi T, Matsui M, Fujita T et al. Structure and expression of a cloned cDNA for human interleukin-2. Nature 1983; 302:305-10.
6. Smith KA. Interleukin-2: Inception, impact, and implications. Science 1985; 240:1169-76.
7. Rosenberg SA, Lotze MT, Muul LM et al. A progress report on the treatment of 157 patients with advanced cancer using lympho kine-activated killer cells and interleukin-2 or high dose interleukin-2 alone. N Engl J Med 1987; 316:889-97.
8. Tsudo M, Uchiyama T, Uchino H. Expression of Tac antigen on activated normal human B cells. J Exp Med 1984; 160:612-7.
9. Espinoza-Delgado I, Ortaldo JR, Winkler-Pickett R et al. Expression and role of p75 interleukin 2 receptor on human monocytes J Exp Med 1990; 171:1821-6.
10. Weiss A, Imboden JB. Cell surface molecules and early events involved in human T lymphocyte activation. Adv Immunol 1987; 41:1-38.
11. Ullman KS Northrop JP, Verweij CL et al. Transmission of signals from the T lymphocyte antigen receptor to the genes responsible for cell proliferation and immune function: the missing link. Ann Rev Immunol 1990; 8:421-52.
12. Durand DB, Ruxh MR, Morgan JG et al. A 275-base pair frag-

ment at the 5' end of the interleukin 2 gene enhances expression from a heterologous promoter in response to signals from the T-cell antigen receptor. J Exp Med 1987; 165:395-407.
13. Fujita T, Shibuya H, Ohashi T et al. Regulation of human IL-2 gene: functional DNA sequences in the 5' flanking region for the gene expression in activated T lymphocytes. Cell 1986; 46:401-7.
14. Ullman KS, Flangan WM, Edwards CA et al. Activation of early gene expression in T lymphocytes by Oct-1 and inducible protein, OAP40. Science 1991; 254:558-62.
15. Jain J, Valge-Archer VE, Rao A. Analysis of the AP-1 sites in the IL-2 promoter. J Immunol 1992; 148:1240-50.
16. Kamps MP, Corcoran L, LeBowitz JH et al. The promoter of the human interleukin-2 gene contains two octamer binding sites and is partially activated by the expression of Oct-2. Mol Cell Biol 1990; 10:5464-72.
17. Nothrop JP, Ho SF, Chen L et al. NF-AT components define a family of transcription factors targeted in T-cell activation. Nature 1994; 369:497-502.
18. Mattila PS, Ullman KS, Fiering S et al. The action of cyclosporin A and FK506 suggest a novel step in the activation of T lmphocytes. EMBO J 1990; 9:4425-33.
19. Fraser JD, Weiss A. Regulation of T cell lymphokine transcription by the accessory molecule CD28. Mol Cell Biol 1992; 12:4357-63.
20. Shaw G, Kamen R. A conserved AU sequence from the 3' untranslated region of GM-CSF mRNA mediates selective mRNA degradation. Cell 1986; 46:659-67.
21. Robb RJ, Munck A, Smith KA. T-cell growth factors: quantification, specificity, and biological relevance. J Exp Med 1981; 154:1455-74.
22. Uchiyama T, Broder S, Waldmann TA. A monoclonal antibody (anti-Tac) reactive with activated and functionally mature human T cells. J Immunol 1981; 126:1393-7.
23. Leonard WJ, Depper JM, Uchiyama T et al. A monoclonal antibody that appears to recognize the receptor for human T-cell growth factor, partial characterization of the receptor. Nature 1982; 300:267-9.
24. Robb RJ, Greene WC. Direct demonstration of the identity of T cell growth factor binding protein and the Tac antigen. J Exp Med 1983; 158:1332-7.
25. Nikaido T, Shimizu A, Ishida N et al. Molecular cloning of cDNA encoding human interleukin 2 receptor. Nature 1984; 311:631-5.
26. Leonard WJ, Depper JM, Crabtree GR et al. Molecular cloning and expression of cDNAs for the human interleukin-2 receptor. Nature 1984; 311:626-31.
27. Sharon M, Klausner RD, Cullen BR et al. Novel interleukin-2 receptor subunit detected by cross-linking under high-affinity condi-

tions. 1986:859-863.
28. Tsudo M, Kozak RW, Goldman CK et al. Demonstration of non-Tac peptide that binds interleukin-2: A potential participant in a multichain interleukin-2 receptor complex. Proc Natl Acad Sci USA 1986; 83:9694-8.
29. Hatakeyama M, Tsudo M, Minamoto S et al. Interleukin-2 receptor β chain gene: generation of three forms by cloned human α and β chain cDNAs. Science 1989; 244:551-6.
30. Waldmann T. The interleukin-2 receptor. J Biol Chem 1991; 266:2681-4.
31. Takeshita T, Ohtani K, Asao H et al. An associated molecule, p64, with IL-2 receptor β chain. Its possible involvement in the formation of the functional intermediate-affinity IL-2 receptor complex. J Immunol 1992; 148:2154-8.
32. Takeshita T, Asao H, Ishii N et al. Cloning of the γ chain of the human IL-2 receptor. Science 1992; 257:379-82.
33. Arima N, Kamio M, Imada K et al. Pseudo-high affinity interleukin-2 (IL-2) receptor lacks the third component that is essential for functional IL-2 binding and signaling. J Exp Med 1992; 176:1265-72.
34. Leonard WJ, Depper JM, Kanehisa M et al. Structure of the human interleukin-2 receptor gene. Science 1985; 230:633-9.
35. Bohnlein E, Lowenthal JW, Siekevitz M et al. The same inducible nuclear proteins regulates mitogen activation of both the interleukin-2 receptor- α gene and type 1 HIV. Cell 1988; 53:827-36.
36. Greene WC, Bohnlein E, Ballard DW. HIV-1, HTLV-I and normal T-cell growth: transcriptional strategies and surprises. Immunol Today 1989; 10:272-8.
37. Hatakeyama M, Mori H, Doi T et al. A restricted cytoplasmic region of IL-2 receptor β chain is essential for growth signal transduction but not for ligand binding and internalization. Cell 1989; 59:837-45.
38. Shibuya H, Yoneyama M, Ninomiya-Tsuji J et al. IL-2 and EGF receptors stimulate the hematopoietic cell cycle via different signaling pathways: Demonstration of a novel role for c-myc. Cell 1992; 70:57-67.
39. Giri JG, Ahdieh M, Eisenman J et al. Utilization of the β and γ chains of the IL-2 receptor by the novel cytokine IL-15. EMBO J 1994; 13:2822-30.
40. Nakamura Y, Russell SM, Mess SA et al. Heterodimerization of the IL-2 receptor β- and γ-chain cytoplasmic domains is required for signaling. Nature 1994; 369:330-6.
41. Kondo M, Takeshita T, Ishii N et al. Sharing of the interleukin-2 (IL-2) receptor γ chain between receptors for IL-2 and IL-4. Science 1993; 262:1874-7.
42. Noguchi M, Nakamura Y, Russell S et al. Interleukin-2 receptor γ

chain: A functional component of the interleukin-7 receptor. Science 1993; 262:1877-80.
43. Tsudo M, Goldman CK, Bongivanni WC et al. The p75 peptide is the receptor for interleukin 2 expressed on large granular lymphocytes and is responsible for the interleukin 2-activation of these cells. Proc Natl Acad Sci USA 1987; 84:5394-8.
44. Hori T, Uchiyama T, Onishi R et al. Characteristics of the IL-2 receptor expressed on large granular lymphocytes from patients with abnormally expanded large granular lymphocytes. Implication of a non-Tac IL-2-binding peptide. J Immunol 1988; 140:4199-203.
45. Ohashi Y, Takeshita T, Nagata K et al. Differential expression of the IL-2 receptor subunits, p55 and p75 on various populations of primary peripheral blood mononuclear cells. J Immunol 1989; 143:3548-55.
46. Lowenthal JW, Greene WC. Contrasting interleukin 2 binding properties of the (p55) and (p70) protein subunits of the human high-affinity interleukin 2 receptor. J Exp Med 1987; 166:1156-61.
47. Wang HM, Smith KA. The interleukin-2 receptor: functional consequences of its bimolecular structure. J Exp Med 1988; 166:1055-69.
48. Rubin LA, Kurman CC, Fritz ME et al. Soluble interleukin-2 receptors are released from activated human lymphoid cells in vitro. J Immunol 1985; 135:3172-7.
49. Hattori T, Uchiyama T, Toibana T et al. Surface phenotype of Japanese adult T-cell leukemia cells characterized by monoclonal antibodies. Blood 1981; 58:645-7.
50. Uchiyama T, Hattori T, Wano Y et al. Cell surface phenotype and in vitro function of adult T cell leukemia cells. Diag Immunol 1983; 1:150-4.
51. Uchiyama T, Hori T, Tsudo M et al. Interleukin-2 receptor (Tac antigen) expressed on adult T cell leukemia cells. J Clin Invest 1985; 76:446-53.
52. Kodaka T, Uchiyama T, Ishikawa T et al. Interleukin-2 receptor β-chain (p70-75) expressed on leukemic cells from adult T cell leukemia patients. Jpn J Cancer Res 1990; 81:902-8.
53. Waldmann TA, Greene WC, Sarin PS et al. Functional and phenotypic comparison of human T cell leukemia/lymphoma virus, positive adult T cell leukemia with human T cell leukemia/ lymphoma virus, negative Sezary leukemia and their distinction using anti-Tac monoclonal antibody identifying the human receptor for Tcell growth factor. J Clin Invest 1984; 73:1711-8.
54. Depper JM, Lenard WJ, Kronke M et al. Augmented T cell growth factor expression in HTLV-I-infected human leukemic T cells. J Immunol 1984; 133:1691-5.
55. Sugamura K, Fujii M, Kobayashi N et al. Retrovirus-induced expression of interleukin 2 receptors on cells of human B-cell lin-

eage. Proc Natl Acad Sci USA 1984; 81:7441-5.
56. Hoshino S, Oshimi K, Tsudo M et al. Flow cytometric analysis of expression of interleukin-2 receptor β chain (p70-75) on various leukemic cells. Blood 1990; 76:767-74.
57. Sugamura K. Expression of IL-2 receptor subunits, α, β, and γ, on HTLV-I-transformed T cells. In: Takatsuki K, Hinuma Y, Yoshida M ed. Advanced in Adult T Cell Leukemia and HTLV-I Research. Tokyo: Japan Scientific Societies Press, 1992:119-128.
58. Arima N, Daitoku Y, Ohgaki S et al. Autocrine growth of interleukin 2 producing leukemic cells in a patient with adult T cell leukemia. Blood 1986; 68:779-82.
59. Kodaka T, Uchiyama T, Umadome H et al. Expression of cytokine mRNA in leukemic cells from adult T cell leukemia patients. Jpn J Cancer Res 1989; 80:531-6.
60. Inoue J, Seiki M, Taniguchi T et al. Induction of interleukin 2 receptor gene expression by $p40^x$ encoded by human T-cell leukemia virus type I. EMBO J 1986; 5:2883-8.
61. Umadome H, Uchiyama T, Hori T et al. Close association between interleukin-2 receptor mRNA expression and human T cell leukemia/lymphoma virus type I viral RNA expression in short-term-cultured leukemic cells from adult T cell leukemia patients. J Clin Invest 1988; 81:52-61.
62. Seiki M, Hattori S, Hirayama Y et al. Human adult T cell leukemia virus: Complete nucleotide sequence of the provirus genome integrated in leukemia cell DNA. Proc Natl Acad Sci USA 1983; 80:3618-22.
63. Sodroski JG, Rosen CA, Haseltine WA. Trans-acting trascriptional activation of the long terminal repeat of human T lymphotropic viruses in infected cells. Science 1984; 225:381-5.
64. Seiki M, Inoue J, Takeda T et al. Direct evidence that $p40^x$ of human T-cell leukemia virus type I is a trans-acting transcriptional activator. ENBO J 1986; 5:561-5.
65. Siekevitz M, Feinber MB, Holbrook N et al. Activation of interleukin 2 and interleukin 2 receptor (Tac) promoter expression by the trans-activator (tat) gene product of human T-cell leukemia virus type I. Proc Natl Acad Sci USA 1987; 84:5389-93.
66. Ballard DW, Bohnlein E, Lowenthal JW et al. HTLV-I tax induces cellular proteins that activate the κB element in the IL-2 receptor α gene. Science 1988; 241:1652-5.
67. Leung K, Nabel GJ. HTLV-1 transactivator induces interleukin-2 receptor expression through an NF-κB-like factor. Nature 1988; 333:776-8.
68. Hirai H, Suzuki T, Fujisawa J et al. Tax protein of human T-cell leukemia virus type I binds to the ankirin motifs of inhibitory factor κB and induces nuclear translocation of transcription factor NF-κB proteins for transcriptional activation. Proc Natl Acad Sci

USA 1994; 91:3584-8.
69. Rubin LA, Nelson DL. The soluble interleukin-2 receptor: biology, function, and clinical application. Ann Intern Med 1990; 113:619-27.
70. Marcon L, Rubin LA, Kurman CC et al. Elevated serum levels of soluble Tac peptide in adult T-cell leukemia: Correlation with clinical status during chemotherapy. Ann Intern Med 1988; 15:274-9.
71. Motoi T, Uchiyama T, Uchino H et al. Serum soleuble interleukin-2 receptor levels in patients with adult T-cell leukemia and human T-cell leukemia/lymphoma virus type-I seropositive healthy carriers. Jpn J Cancer Res (Gann) 1988; 79:593-9.
72. Yasuda N, Lai PK, Ip SH et al. Soluble IL-2 receptors in sera of Japanese patients with adult T cell leukemia mark activity of disease. Blood 1988; 71:1021-6.
73. Rubin LA, Jay G, Nelson DL. The released interleukin 2 receptor binds interleukin 2 efficiently. J Immunol 1986; 137:381-4.
74. Miyatake S, Seiki M, Yoshida M et al. T-cell activation signals and human T-cell leukemia virus type I-encoded p40^x protein activate the mouse granulocyte-macrophage colony-stimulating factor gene through a common DNA element. Mol Cell Biol 1988; 8:5581-7.
75. Tschachler E, Robert-Guroff M, Gallo RC et al. Human T-lymphotropic virus I-infected T cells constitutively express lymphotoxin in vitro. Blood 1989; 73:194-201.
76. Kim SJ, Kehrl JH, Burton J et al. Transactivation of the transforming growth factor β1 (TGF-β1) gene by human T lymphotropic virus type 1 tax: a potential mechanism for the increased production of TGF-β1 in adult T cell leukemia. J Exp Med 1990; 172:121-9.
77. Fujii M, Niki T, Mori T et al. HTLV-I tax induces expression of various immediate early serum response genes. Oncogene 1991; 6:1023-9.
78. Sakamoto KM, Nimer SD, Rosenblatt JD et al. HTLV-I and HTLV-II tax transactivate the human EGR-1 promoter through different cis-acting sequences. Oncogene 1992; 7:2125-30.
79. Jeang KT, Widen SG, Semmers IV OJ et al. HTLV-I transactivator protein, Tax, is a transrepressor of the human β polymerase gene. Science 1990; 247:1082-4.

= CHAPTER 6 =

PROLIFERATION OF ATL CELLS IN VIVO

Takashi Uchiyama

Subsequent to the discovery of HTLV-I, seroepidemiological and virological studies were performed to clarify the relationship between ATL and HTLV-I, and the etiologic role of HTLV-I was established.[1-6] Several epidemiological facts should be taken into account when we consider the mechanism of oncogenesis in ATL: (1) ATL develops after a long latent period from HTLV-I infection; and (2) very few people infected with HTLV-I actually develop leukemia.[7] Stochastic analysis of the age-specific occurrence of ATL in 357 cases showed that a simple Weibull distribution function fit well as a model. In addition, the mode of ATL onset was log-linear in this model and the curves for males and females overlapped completely.[8] Furthermore, it suggested the presence of age-dependent accumulation of leukemogenic events which were estimated to be five within HTLV-I-infected T cells prior to the development of ATL.[8] Indeed, the low incidence of ATL among HTLV-I-infected carriers and the long latent period suggests that several sequential changes may be needed for full development of ATL (Fig. 6.1). Abnormal expression of p53 suppressor oncogene due to point mutation or deletion, which has been demonstrated in some ATL cases,[9,10] occurs in the advanced stage of the development of ATL and may be one of these events.

It is thought that HTLV-I infection triggers an initial key event in virus-infected cells, is followed by an as yet undetermined series of changes, and eventually leads to full development of ATL. The pX region of HTLV-I contains two genes, tax and rex, which regulate

Adult T Cell Leukemia and Related Diseases, edited by Takashi Uchiyama and Junji Yodoi. © 1995 R.G. Landes Company.

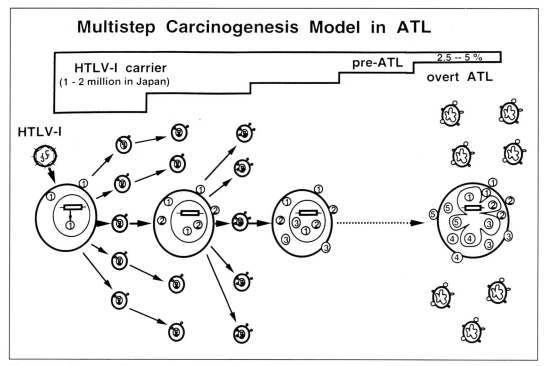

Fig. 6.1. Multistep carcinogenesis model in ATL.

viral replication.[11-13] Because of its activity to enhance the transcription of substantial number of cellular genes, its cell-transforming activity in vitro and its tumorigenic potential demonstrated in pX-transgenic mice, Tax, a 40 kD product of tax gene, is thought to be involved in cell transformation and leukemogenesis.[14-21] It has been demonstrated that many cellular genes are activated or suppressed by HTLV-I infection. It is thought that excessively or scarcely produced protein products may be responsible, in part, for both the formation and modulation of clinical features and, more importantly, neoplastic cell growth (Fig. 6.2). Tax enhances the transcription of cellular genes encoding cytokine and cytokine receptors such as IL-1,[22] IL-2,[15] IL-2R α chain,[14,15,23,24] GM-CSF,[16] TNF-β,[25] TGF-β,[26,27] PTHrP[28] and transcription factors such as c-FOS and EGR.[29] Tax exerts its transcription-enhancing activity probably through the formation of complex together with cellular transcription factors including NF-κB and SRF. On the other hand, it has also been reported that Tax suppresses the

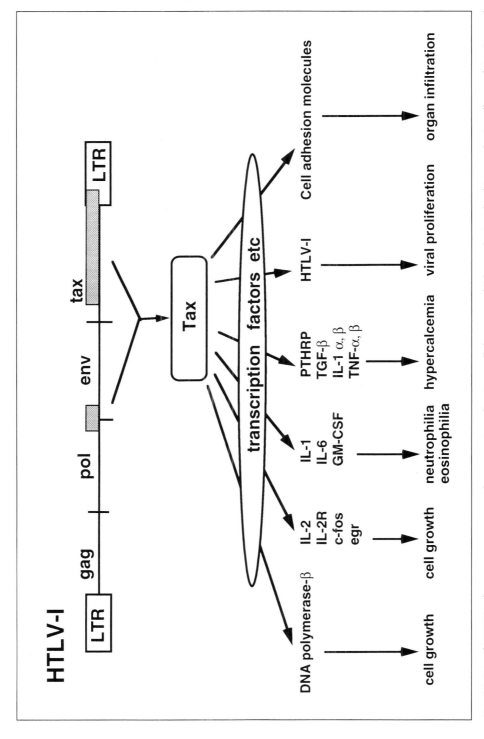

Fig. 6.2. Tax of HTLV-I enhances or suppresses the expression of various cellular genes which may be involved in neoplastic cell growth in vivo and modulate the clinical manifestations in ATL.

transcription of DNA polymerase β,[30] which is involved in the DNA repair mechanism. Some of these cell growth- or death-associated genes, which are abnormally expressed in HTLV-I-infected cells, may play a key role in the leukemogenesis of ATL.

We have been trying to clarify the mechanism of neoplastic cell growth in ATL by focusing on the abnormal expression of IL-2R. Overexpression of IL-2R α chain is one of the characteristics of HTLV-I-infected T cells. Virtually all HTLV-I-infected T cell lines established from ATL patients or HTLV-I carriers, as well as the majority of ATL leukemic cells, constitutively express IL-2R α chain, both at mRNA and protein levels.[31-33] It has been reported that the serum level of soluble IL-2R α chain released from ATL cells was markedly elevated in patients[34,35] and that expression of IL-2R α chain mRNA was closely associated with the expression of HTLV-I viral RNA in short-term cultured leukemic cells from ATL patients.[33] Furthermore, it was clearly demonstrated that the introduction of pX of HTLV-I into cells induced expression of IL-2R α chain and IL-2.[12,23,24] However, the precise role of abnormally expressed IL-2R in the development of ATL as well as neoplastic cell growth remains obscure. Leukemic T cells isolated from the majority of ATL patients do not proliferate in vitro in response to exogenous IL-2 and many HTLV-I-infected T cell lines do not require IL-2 for cell growth, although their culture is usually initiated in the presence of IL-2. In addition, it has been reported that IL-2 mRNA was not detectable in peripheral blood leukemic cells from ATL patients by Northern blot analysis.[36] These results may indicate that the proportion of ATL cells capable of proliferating by an IL-2 autocrine or paracrine mechanism is too small to be detected by usual assays. These results may also indicate that abnormally expressed IL-2R may play a key role at certain stages of cell transformation and ATL development, but is then no longer involved in neoplastic cell proliferation in overt ATL.

An alternative possibility is that the IL-2R system dysregulated in ATL may continue to produce a second signal without IL-2 binding to IL-2R, which leads to neoplastic cell growth. Most experiments which have addressed these critical questions have been performed in in vitro systems. Although these studies have not clarified the mechanism of neoplastic cell proliferation or leukemogenesis, they have provided important clues in this area.

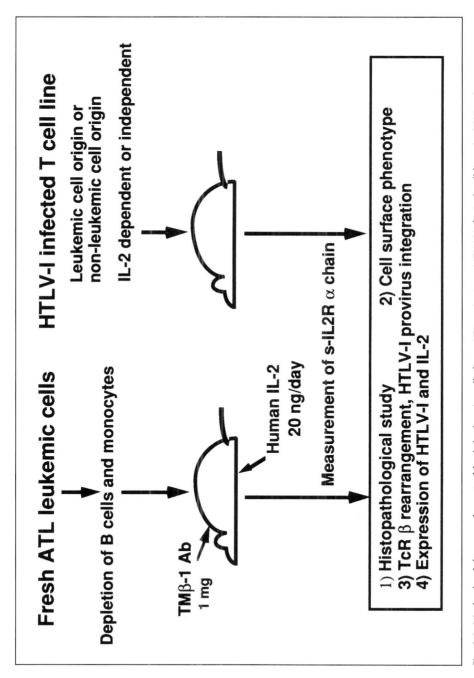

Fig. 6.3. Methods of the engraftment of fresh leukemic cells from ATL patients or HTLV-I-infected cell lines in SCID mice.

Table 6.1. Engraftment of fresh leukemic cells from ATL patients (summary)

Case	Type	Cell Origin	Tumor Growth	Elevation of s-Tac	Survival (Days)	PB	LN	Organ Infiltration Spleen	Liver	Lung	Kidney
1	L	LN	Yes	Yes	47	–	+	++	+++	+++	+++
2	C	PB	Yes	Yes	38	–	+++	++	–	–	–
3	L	LN	Yes	Yes	95	NE	NE	NE	NE	NE	NE
4	C	PB	No	No							
5	C	PB	Yes	Yes	32	+++	+	++	++	++	+
6	A	PB	No	No							
7	L	LN	Yes	Yes	44	NE	NE	++	++	NE	
8	L	LN	Yes	Yes	21	–	+++	+	++	–	–

s-Tac:serum level of the soluble form of human IL-2 receptor α chain(Tac), PB:peripheral blood, LN:lymph node, L: lymphoma type, C:chronic type, A:acute type, NE:not examined; Organ infiltration: – :none, + :slight, ++ :moderate, +++ :marked.

The prognosis of ATL is still very poor[37] as progress in the treatment of ATL has been limited since the first description of the disease. An appropriate model of in vivo proliferation of ATL cells would provide us with more powerful experimental approaches to address these issues. Such a model would provide important information so that we might better understand the mechanism of leukemogenesis and develop an effective treatment for ATL. Recently, we developed an animal model using SCID mice which is useful for study of the mechanism of ATL development and in vivo cell proliferation of ATL cells.[38]

ENGRAFTMENT OF FRESH LEUKEMIC CELLS FROM ATL PATIENTS IN SCID MICE

SCID mice, due to defects in mature normal T and B cells, could successfully be engrafted with human hematopoietic cells or neoplastic cells.[39-42] For example, in vivo tumor formation of Epstein-Barr virus (EBV)-infected human B cells in SCID mice has been reported and used as a model to study the mechanism of B cell lymphomagenesis.[43] We have been trying to make a model of in vivo cell proliferation of leukemic cells from ATL patients using SCID mice. We succeeded in making such a model after failures in earlier studies of SCID mice in which we almost always observed preferential cell growth of EBV-infected B cells from patients. We initially attempted to inject SCID mice with peripheral blood leukemic cells from acute type ATL patients with a very high peripheral blood leukocyte count or lymph node cells obtained by biopsy, or put small pieces of lymph nodes on the upper pole of the mouse kidney. These early trials, however, resulted in the formation of tumors composed mainly of B cells, but not leukemic T cells, 2 or 3 months after cell injection or transplantation.

Successful engraftment of fresh leukemic cells in SCID mice has been brought about by three techniques (Fig. 6.3).[38] First, we injected cells intensively depleted of B cells and monocytes by the immunomagnetic beads method in order to diminish the possibility of B cell tumor growth. Second, we used SCID mice pretreated with anti-murine IL-2R β chain antibody (TMβ1 antibody) which markedly diminished the number and almost completely abrogated the function of NK cells.[44] Third, we injected human

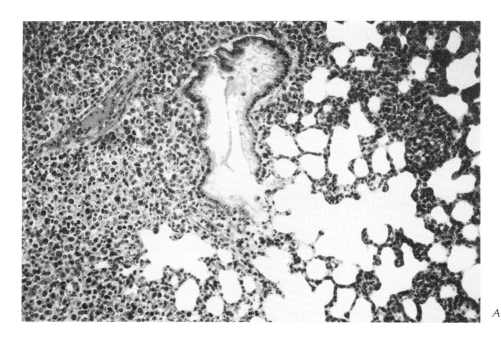

Figs. 6.4A and 6.4B. Leukemic cell infiltration into the lung (A) or kidney (B) of the mouse injected with fresh leukemic cells from an ATL patient (case 1).

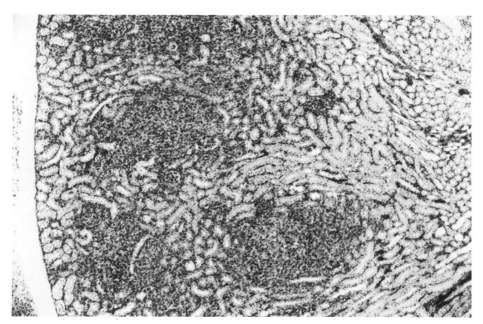

recombinant IL-2 into SCID mice daily to support the cell proliferation of ATL cells in vivo. It remains to be determined whether the administration of IL-2 is requisite to successful engraftment, although it appeared to accelerate the preferential cell growth of ATL cells in vivo in most of the ATL cases studied. In addition, we measured the serum levels of the soluble form of human IL-2R α chain(Tac) in mice by a sandwich ELISA method, which turned out to be a very useful indicator to monitor the number of HTLV-I-infected cells in mice as reported in ATL cases.

We found tumor growth in SCID mice injected with ATL cells from six out of eight ATL patients 5 to 7 weeks after the inoculation of ATL cells (Table 6.1). Infiltration of ATL cells into various organs was observed in four mice injected with cells from four different ATL patients (Figs. 6.4A and 6.4B). We found small tumors at the root of mesentery and enlargement of lymphoid organs (spleen, thymus and lymph nodes) at autopsy. Two mice became leukemic (Fig. 6.5) and peripheral blood white blood cell count increased to the level of 30×10^9/L. Histological examination revealed

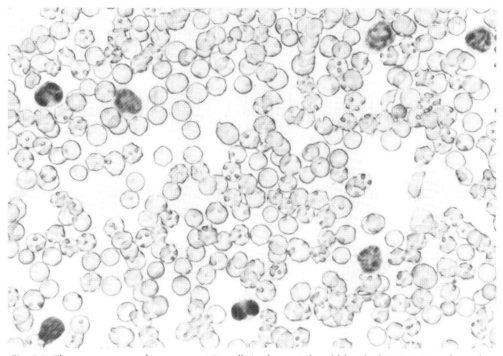

Fig. 6.5. The appearance of numerous ATL cells in the peripheral blood of SCID mice where the number of murine lymphocytes is usually very low.

the infiltration of atypical lymphoid cells which were reactive with human CD3 and UCHL1(CD45RO) antibodies, but not with L26(CD20) mAb into portal regions of the liver, interstitial regions of the lungs, and the kidney. The cell infiltration predominantly into lymphoid organs, as seen in lymphoma type ATL, was observed in two mice injected with cells from two different ATL patients. The enlarged lymph nodes showed the histological characteristics of a pleomorphic type lymphoma with scattered human $CD3^+$ giant cells, which is often observed in the lymph nodes of ATL patients. Consistent with these findings, the spleen of the mouse showed the infiltration of human $CD3^+$ cells mainly into the white pulp region. However, the infiltration of the lymphoid cells into other organs including the liver, kidneys and lungs was minimal in these two mice.

We next studied the cell surface phenotype of the cells proliferating and forming the tumor in SCID mice by flow cytometric analysis after staining cells with various monoclonal antibodies. The major population of the cells analyzed were reactive with anti-human CD3, CD4 and IL-2R α chain antibodies, which was the same phenotype as that of the leukemic cells from original ATL patients.

It was confirmed by Southern blot hybridization analysis using pX of HTLV-I and Cβ1 of T cell receptor β chain (TcRβ) gene, taking advantage of a random integration of HTLV-I provirus into each ATL cell clone and a different TcRβ gene rearrangement in each T cell clone that the cells proliferating in SCID mice were derived from the original leukemic cell clone. We detected the same integration site of HTLV-I and the same TcRβ gene rearrangement pattern as those of the original leukemic cells in all four cases examined, clearly confirming the clonal identity (Fig. 6.6). We also detected additional clear bands on the filter in one mice. This indicated cell proliferation of oligoclones other than the ATL cell clone, which was coincident with the detection of $CD3^+CD8^+$ cell population by flow cytometric analysis. Exogenous human IL-2 injected into mice may promote the cell growth of HTLV-I-infected nonleukemic cells expressing IL-2R α chain as well as leukemic cells in vivo in some cases. The events occurring in the selection of certain HTLV-I-infected T cell clone(s) in SCID mice might reflect key changes during early stages of the development of ATL. Monoclonally-expanded HTLV-I-infected T cells have been detected in a small proportion of HTLV-I-infected individuals who did not show sufficient symptoms or signs for a definite diagnosis of ATL.[45] A minority of such individuals will eventually develop overt ATL. The model using SCID mice may display such processes to us in a short period by condensing some stages of the natural course of leukemogenesis initiated by HTLV-I-infection.

We unexpectedly found that leukemic cells from peripheral blood, as well as lymph nodes, could proliferate and form tumors in the mice. We considered before the study that the lymph nodes, not the peripheral blood of ATL patients, may be one of the organs where leukemic cells actively proliferate since spontaneous ^{3}H-thymidine incorporation by fresh peripheral blood leukemic cells was low in most of the ATL cases studied and cell proliferation-associated molecules were more commonly expressed in lymph node cells. The dissociation between the results of an in vitro proliferation assay and those of an in vivo tumorigenicity study in SCID mice not only illustrates the importance of in vivo studies but also alludes to complex mechanisms underlying ATL abnormal cell growth in vivo.

CELL GROWTH OF HTLV-I-INFECTED CELL LINES IN SCID MICE

We next tried to see whether HTLV-I-infected cell lines, immortalized with or without IL-2 in vitro could grow in vivo to kill SCID mice. Seven HTLV-I-infected cell lines were tested. Three cell lines were derived from the original leukemic cell clone, which was determined by the same integration site of HTLV-I provirus and the same pattern of T cell receptor β chain gene rearrangement between them. Three cell lines have been maintained with human recombinant IL-2 and the remaining four cell lines grow without IL-2 (IL-2 independent). The successful engraftment of the injected cell lines were observed in four cell lines (Figs. 6.5A and 6.5B) and three other cell lines injected into SCID mice could not grow and did not show any tumorigenicity. It appears that HTLV-I-infected cell lines derived from the original leukemic cells are easily engrafted, while HTLV-I-infected cell lines derived from nonleukemic cell clones are hardly engrafted. These results suggest that in vitro immortalized cell lines derived from the leukemic cells but not nonleukemic cell lines may acquire

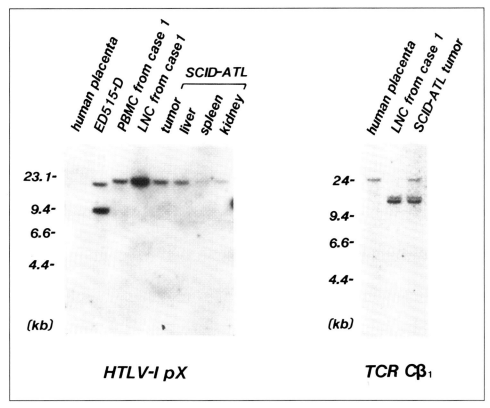

Fig. 6.6. Southern blot hybridization of DNA obtained from peripheral blood leukemic cells (PBMC) and lymph node cells (LN) from patient 1, and cells from murine tumor, liver, spleen and kidney (SCID-ATL). The probes used are pX of HTLV-I and Cβ1 of T cell receptor β chain gene (TCR Cβ1). ED515-D is a HTLV-I-infected cell line used as a positive control.

enough changes to promote cell proliferation in SCID mice and ultimately result in the death of the mice (Fig. 6.6).

Another interesting finding obtained from the study is that cell lines growing in an IL-2 dependent manner in vitro could proliferate in vivo without the supply of exogenous human IL-2. IL-2 binding to its receptor induces the activation of cyclins/cdks which leads to the progression of competent factor-activated T cells at early G1 phase to late G1/S phase. Considering no production of IL-2 in SCID mice and no biological activity of murine IL-2 for human T cells expressing human IL-2R, it would be interesting to determine what makes injected cells progress from G1 to S phase of cell cycle instead of IL-2 in SCID mice.

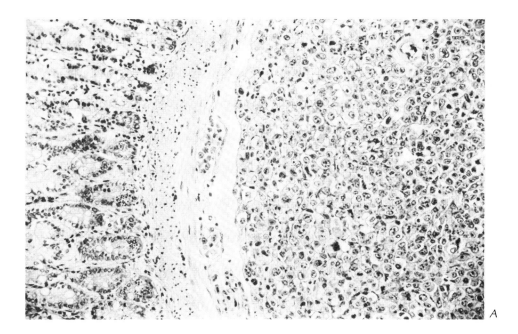

Fig. 6.7. Infiltration of ATL-43T cell line into a mesenteric lymph node(A) and spleen (B) of the SCID mouse.

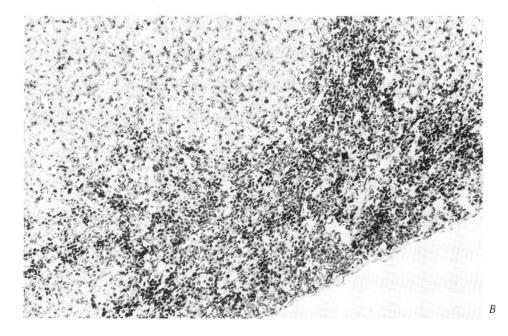

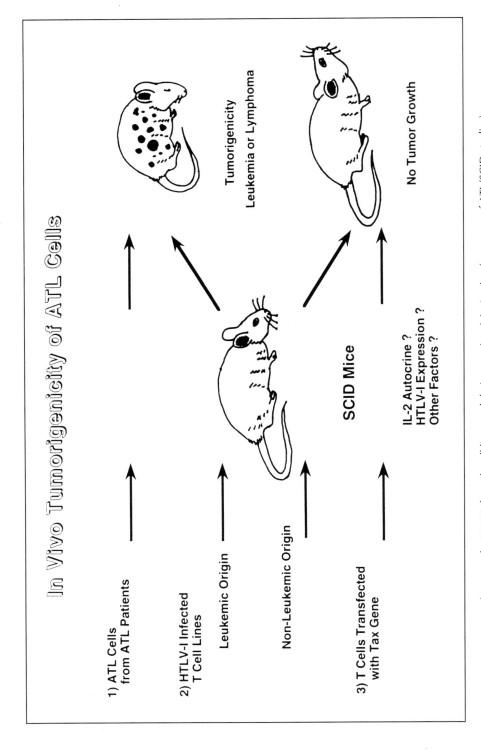

Fig. 6.8. *In vitro immortalization of HTLV-I-infected cell lines and their tumorigenicity in vivo (summary of ATL/SCID studies).*

HTLV-I expression has been hardly detectable in fresh leukemic cells and most of the HTLV-I-infected cell lines although it has been agreed that HTLV-I is a causative virus of ATL. Furthermore, it has been demonstrated that HTLV-I Tax protein transactivates a variety of cellular genes including cell proliferation- or cell death-associated molecule genes. These molecules which are abnormally expressed through the effect of Tax protein may be deeply involved in the neoplastic cell growth in vivo in ATL. It is, therefore, still controversial that the expression of HTLV-I is essential and prerequisite for the neoplastic cell growth in ATL. The expression of HTLV-I was examined by Northern blot hybridization and reverse transcription polymerase chain reaction (RT-PCR) method. Any cell lines except for one cell line cultured in vitro did not express HTLV-I viral RNA. Three cell lines which were successfully engrafted in SCID mice did not express viral RNA in vivo either. Thus, it appears that the expression of HTLV-I is not required for the proliferation of the cells at least in SCID mice.

Regarding the tumorigenicity of HTLV-I-infected and IL-2-dependent cell lines in SCID mice, we examined the production of human IL-2 by the cell line cells which had been injected into mice and proliferated in vivo to see whether IL-2 autocrine mechanism may begin to work in SCID mice. Northern blot hybridization analysis, however, did not show any human IL-2 mRNA expression in cells growing in SCID mice. It was also confirmed by the finding that no signal of IL-2 mRNA was detected in the cells recovered from the mice by a more sensitive RT-PCR technique. It is, therefore, most unlikely that HTLV-I-infected cell lines proliferate by an ordinary IL-2 autocrine mechanism.

The SCID mouse/ATL model we developed appears to help us to distinguish the immortalization in vitro and the tumorigenicity in vivo of HTLV-I-infected cells and, therefore, to explore the factor(s) which determines the essential nature of neoplastic cells.

RADIOMETRIC STUDIES ON THE PROLIFERATION OF ATL-43T CELLS IN VIVO

A characteristic clinical manifestation of ATL is infiltration of leukemic cells into various organs such as lymph nodes, spleen, lungs, kidneys and skin. We do not know in exactly which organs the leukemic cells proliferate, although lymph nodes are suspected to be the primary organs of leukemic cell proliferation. We studied the sites of cell proliferation and infiltration in an ATL/SCID model using radiometric analysis. ATL-43T cells derived from the original leukemic cell clone and grown IL-2 dependently in vitro was found to be tumorigenic without exogenous IL-2 in SCID mice.

We traced ATL-43T cells in SCID mice using ^{111}In-oxine labeled ATL-43T cells, ^{125}I-labeled anti-Tac antibody or ^{111}In-labeled anti-Tac antibody. The studies using ATL-43T cells labeled with ^{111}In, which has a short half life, were able to disclose the early (up to 1 week after cell injection) distribution of the cells. On day 2, we could not detect significant accumulation of radioactivity in any organs, although a slight elevation was detectable in the spleen and thymus. On day 6, we detected a significantly higher accumulation of radioactivity in the spleen and thymus of SCID mice, compared with that of the same organs of control CB17 mice. The injection of ^{125}I- or ^{111}In-labeled anti-Tac antibody, followed by measurement of organ radioactivity enabled us to trace the in vivo cell growth of ATL-43T cells expressing IL-2R α chain from the injection of the cells to the sacrifice of the mice. A significant accumulation of ^{125}I-anti-Tac antibody was detectable in the spleen as early as 2 weeks after cell inoculation, while that of ^{111}In-labeled anti-Tac antibody in the spleen was undetectable. It is probably because ^{111}In is preferentially trapped in the spleen and liver. Four weeks after the intraperitoneal injection of ATL-43T cells the accumulation of the radioactivity by ^{125}I- or ^{111}In-labeled anti-Tac antibody in the spleen and thymus was more obvious, compared with radiolabeled control RPC antibody with the same isotype as anti-Tac antibody. The increased accumulation of ^{111}In- or ^{125}I-labeled anti-Tac antibody in the spleen and thymus was time-dependent. Serial histopathological examinations of the SCID mice engrafted with ATL-43T cells confirmed that the cells preferentially moved to and proliferated in the thymus and spleen of SCID mice. Accumulation of radiolabeled anti-Tac antibody in organs where the infiltration of ATL-43T cell was demonstrated by histopathological examinations strongly supports efficient targeting of ATL cells in vivo by anti-Tac antibody. Indeed, Waldmann et al published promising data concerning therapeutic approaches utilizing unconjugated or radiolabeled anti-Tac antibody in ATL.[46,47]

Thus the radiometric analysis of the organ distribution of ATL cells using an ATL/SCID model has provided us with useful information for understanding the nature of the cell growth of ATL cells in vivo and for obtaining an experimental basis for new therapeutic approaches.

ACKNOWLEDGMENTS

The review article described in this chapter is based on collaborative work with Drs. Kazunori Imada, Akifumi Takaori-Kondo, Michiyuki Maeda, Masayuki Miyasaka, Kunitada Shimotohno and Makoto Hosono, to whom the author, Takashi Uchiyama, is very grateful.

REFERENCES

1. Poiesz BJ, Ruscetti FW, Gazdar AF et al. Detection and isolation of type-C retrovirus particles from fresh and cultured lymphocytes of a patient with cutaneous T-cell lymphoma. Proc Natl Acad Sci USA 1980; 77:7415-9.
2. Hinuma Y, Nagata K, Hanaoka M et al. Antigen in an ATL cell line and detection of antibodies to the antigen in human sera. Proc Natl Acad Sci U.S. 1981; 78:6476-80.
3. Yoshida M, Miyoshi I, Hinuma Y. Isolation and characterization of retrovirus from cell lines of human adult T cell leukemia and its implication in the disease. Proc Natl Acad Sci USA 1982; 79:2031-5.
4. Hinuma Y, Komoda H, Chosa T et al. Antibodies to adult T-cell leukemia virus-associated antigen (ATLA) in sera from patients with ATL and controls in Japan: a nationwide sero-epidemiologic study. Int J Cancer 1982; 29:631-5.
5. Blattner WA, Kalyanaraman VS, Robert-Guroff M et al. The human type-C retrovirus HTLV in blacks from the Caribbean region and relationship to adult T-cell leukemia/lymphoma. Int J Cancer 1982; 30:257-64.
6. Yoshida M, Seiki M, Yamaguchi K et al. Monoclonal integration of human T-cell leukemia provirus in all primary tumors of adult T-cell leukemia suggests causative role of human T-cell leukemia virus in disease. Proc Nat Acad Sci USA 1984; 81:2534-7.
7. Tajima K, The T- and B-cell Malignancy Study Group. The fourth nationwide study of adult T-cell leukemia/lymphoma (ATL in Japan): Estimates of risk of ATL and its geographical and clinical features. Int J Cancer 1990; 45:237-43.

8. Okamoto T, Ohno Y, Tsugane S et al. Multi-step carcinogenesis model for adult T-cell leukemia. Jpn J Cancer Res 1989; 80:191-5.
9. Cesarman C, Chadburn A, Inghirami G et al. Structural and functional analysis of oncogenes and tumor suppressor genes in adult T-cell leukemia/lymphoma shows frequent p53 mutations. Blood 1992; 80:3205-16.
10. Sakashita T, Hattori T, Miller CW et al. Mutations of the p53 gene in adult T-cell leukemia. Blood 1992; 79:477-80.
11. Seiki M, Inoue J, Takeda T et al. The $p40^x$ of human T-cell leukemia virus type I is a trans-acting activator of viral gene transcription. Gann 1985; 76:1127-31.
12. Inoue J, Yoshida M, Seiki M. Transcriptional ($p40^x$) and post-transcriptional ($p27^x$-III) regulators are required for the expression and replication of human T cell leukemia virus type I genes. Proc Natl Acad Sci USA 1987; 84:3653-57.
13. Inoue J, Itoh M, Akizawa T et al. HTLV-I rex protein induces accumulation of unspliced RNA in nucleus as well as in cytoplasm. Oncogene 1991; 6:1753-7.
14. Inoue J, Seiki M, Taniguchi T et al. Induction of interleukin 2 receptor gene expression by $p40^x$ encoded by human T-cell leukemia virus type I. EMBO J 1986; 5:2883-8.
15. Siekevitz M, Feinber MB, Holbrook N et al. Activation of interleukin 2 and interleukin 2 receptor (Tac) promoter expression by the trans-activator (tat) gene product of human T-cell leukemia virus type I. Proc Natl Acad Sci USA 1987; 84:5389-93.
16. Miyatake S, Seiki M, Malefijt RDW et al. Activation of T cell-derived lymphokine genes in T cells and fibroblasts; effects of human T cell leukemia virus type I $p40^x$ protein and bovine papilloma virus encoded E2 protein. Nucl Acids Res 1988; 16:6547-66.
17. Grassmann R, Degler C, Muller-Fleckenstein I et al. Transformation to continuous growth of primary human T lymphocytes by human T-cell leukemia virus type I X-region genes transduced by a *Herpesvirus saimiri* vector. Proc Natl Acad Sci USA 1989; 86:3351-5.
18. Akagi T, Shimotohno K. Proliferative response of Tax1-transduced primary human T cells to anti-CD3 antibody stimulation by an interleukin-2-independent pathway. J Virol 1993; 67:1211-7.
19. Tanaka A, Takahashi C, Yamaoka S et al. Oncogenic transformation by the tax gene of the human T-cell leukemia virus type I in vitro. Proc Natl Acad Sci USA 1990; 87:1071-5.
20. Hinrichs SH, Nerenberg M, Reynolds RK et al. A transgenic mouse model for human neurofibromatosis. Science 1987; 237:1340-3.
21. Benvenisty N, Ornitz DM, Bennett GM et al. Brain tumors and lymphomas in transgenic mice that carry HTLV-I LTR/c-myc and Ig/tax genes. Oncogene 1992; 7:2399-405.
22. Wano Y, Feinberg M, Hosking J et al. Stable expression of the tax

gene of type I human T-cell leukemia virus in human T cells activates specific cellular genes involved in growth. Proc Natl Acad Sci USA 1988; 85:9733-7.
23. Leung K, Nabel GJ. HTLV-I transactivator induces interleukin-2 receptor expression through an NF-κB-like factor. Nature 1988; 333:776-8.
24. Ballard DW, Bohnlein E, Lowenthal JW et al. HTLV-I tax induces cellular proteins that activate the κB element in the IL-2 receptor α gene. Science 1988; 241:1652-5.
25. Tschachler E, Bohnlein E, Felzmann S et al. Human T-lymphotropic virus type I *tax* regulates the expression of the human lymphotoxin gene. Blood 1993; 81:95-100.
26. Niitsu Y, Urushizaki Y, Koshida Y et al. Expression of TGF-β gene in adult T cell leukemia. Blood 1988; 71:263-6.
27. Kim SJ, Kehrl JH, Burton J et al. Transactivation of the transforming growth factor β1 (TGF-β1) gene by human T lymphotropic virus type 1 tax: a potential mechanism for the increased production of TGF-β1 in adult T cell leukemia. J Exp Med 1990; 172:121-9.
28. Watanabe T, Yamaguchi K, Takatsuki K et al. Constitutive expression of parathyroid hormone-related protein gene in human T cell leukemia virus type I (HTLV-I) carriers and adult T cell leukemia patients that can be trans-activated by HTLV-I tax gene. J Exp Med 1990; 172:759-65.
29. Fujii M, Niki T, Mori T et al. HTLV-1 tax induces expression of various immediate early serum responsive genes. Oncogene 1991; 6:1023-9.
30. Jeang KT, Widen SG, Semmes IV OJ et al. HTLV-I trans-activator protein, Tax, is a transrepressor of the human β-polymerase gene. Science 1990; 247:1082-4.
31. Uchiyama T, Hori T, Tsudo M et al. Interleukin-2 receptor (Tac antigen) expressed on adult T cell leukemia cells. J Clin Invest 1985; 76:446-53.
32. Yodoi J, Uchiyama T. IL-2 receptor dysfunction and adult T-cell leukemia. Immunol Rev 1986; 92:135-56.
33. Umadome H, Uchiyama T, Hori T et al. Close association between interleukin-2 receptor mRNA expression and human T cell leukemia/lymphoma virus type I viral RNA expression in short-term-cultured leukemic cells from adult T cell leukemia patients. J Clin Invest 1988; 81:52-61.
34. Marcon L, Rubin LA, Kurman CC et al. Elevated serum levels of soluble Tac peptide in adult T-cell leukemia: Correlation with clinical status during chemotherapy. Ann Intern Med 1988; 15:274-9.
35. Motoi T, Uchiyama T, Uchino H et al. Serum soluble interleukin-2 receptor levels in patients with adult T-cell leukemia and human T-cell leukemia/lymphoma virus type-I seropositive healthy carriers. Jpn J Cancer Res (Gann) 1988; 79:593-9.

36. Kodaka T, Uchiyama T, Umadome H et al. Expression of cytokine mRNA in leukemic cells from adult T cell leukemia patients. Jpn J Cancer Res 1989; 80:531-6.
37. Shimoyama M. Treatment of patients with adult T-cell leukemia/lymphoma: overview. In: Takatsuki K, Hinuma Y, Yoshida M ed. Advances in Adult T-cell Leukemia and HTLV-I Research. Japan Scientific Societes Press, Tokyo: 1992; 43-56.
38. Kondo A, Imada K, Hattori T et al. A model of in vivo cell proliferation of adult T cell leukemia. Blood 1993; 82:2501-9.
39. Bosma GC, Custer RP, Bosma MJ. A severe combined immunodeficiency mutation in the mouse. Nature 1983; 301:527-30.
40. Custer RP, Bosma GC, Bosma MJ. Severe combined immunodeficiency (SCID) in the mouse: pathology, reconstitution, neoplasms. Am J Pathol 1985; 120:464-77.
41. McCune JM, Namikawa R, Kaneshima H et al. The SCID-hu mouse: murine model for the analysis of human hematolymphoid differentiation and function. Science 1988; 241:1632-9.
42. Kamel-Reid S, Letarte M, Sirard C et al. A model of human lymphoblastic leukemia in immune-deficient SCID mice. Science 1989; 246:1159-600.
43. Rowe M, Young LS, Crocker J et al. Epstein-Barr virus (EBV)-associated lymphoproliferative disease in the SCID mouse model: implications for the pathogenesis of EBV-positive lymphomas in man. J Exp Med 1991; 173:147-58.
44. Tanaka T, Kitamura F, Nagasaka K et al. Selective long-term elimination of natural killer cells in vivo by an anti-interleukin 2 receptor β chain monoclonal antibody in mice. J Exp Med 1993; 178:1103-7.
45. Ikeda S, Momita S, Kinoshita K et al. Clinical course of human T-lymphotropic virus type I carriers with molecularly detectable monoclonal proliferation of T lymphocytes: defining a low- and high-risk population. Blood 1993; 82:2017-24.
46. Waldmann TA, Ira HP, Otta A et al. The multi-chain interleukin-2 receptor: a target for immunotherapy. Ann Intern Med 1992; 116:148-60.
47. Waldmann TA et al. The interleukin-2 receptor—a target for monoclonal antibody treatment of human T-cell lymphotrophic virus I-induced adult T-cell leukemia. Blood 1993; 82:1701-12.

CHAPTER 7

REDOX REGULATION AND ADF

Junji Yodoi

ATL-DERIVED FACTOR(ADF)/HUMAN THIOREDOXIN (hTRX)

The active factor(s) inducing IL-2R α chain was tentatively called ATL-derived factor (ADF) and extensively analyzed and purified to get the cDNA for it.[1]

Sequential purification using several steps of chromatography resulted in a purified active component measuring 12-13 kDa. cDNA cloning of ADF revealed that ADF was different from many cytokines. Homology analysis showed that ADF is a human homologue of thioredoxin which Holmgren et al studied in prokaryotes.[2] Thioredoxin is involved in many of the biochemical reaction steps which transfer protons to cysteine residues of target proteins (Fig. 7.1).

Subsequent studies by Tagaya et al showed that recombinant ADF (rADF) enhanced IL-2R α chain expression on T cells and NK cells. Growth promoting activity on these cells was also confirmed in cooperation with the Wakasugis.[3] In some EBV-transformed B cells, IL-2R α chain upregulation and growth promotion were also demonstrated. Mitsui et al reported both a potent reducing activity and protein refolding activity of ADF.[4]

MECHANISM OF CYTOPROTECTION BY ADF/TRX

It is still unclear how extracellular ADF/hTRX showed these various cytokine-like activities.[5,6] There is little evidence for the high affinity specific receptor for ADF/hTRX.

Adult T Cell Leukemia and Related Diseases, edited by Takashi Uchiyama and Junji Yodoi. © 1995 R.G. Landes Company.

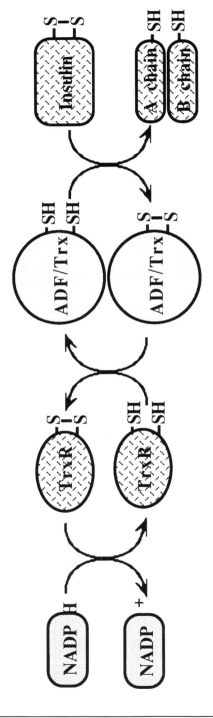

Fig. 7.1. Insulin reducing activity of ADF/thioredoxin system.

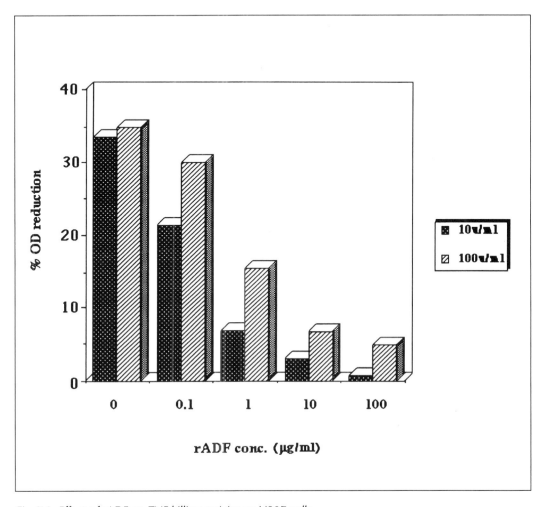

Fig. 7.2. Effect of rADF on TNF killing activity on U937 cells.

Recently, it became clear that extracellular ADF/hTRX can protect the cells against cytotoxicity and oxystress. Matsuda et al showed that cytotoxicity induced by Tumor Necrosis Factor alpha (TNF-α) and anti-Fas monoclonal antibody was effectively protected by rADF (Figs. 7.2, 7.3).[7] In addition, Nakamura et al clearly demonstrated that H_2O_2-induced cytotoxicity of several cell types was also clearly protected by ADF/hTRX. Cytotoxicity by activated neutrophils was also inhibited by ADF/hTRX (Fig. 7.4).[8] It therefore appears likely that redox potential of the cell membrane and intracellular environment is important for protection of the cell membrane, cell growth and viability (Fig. 7.5).

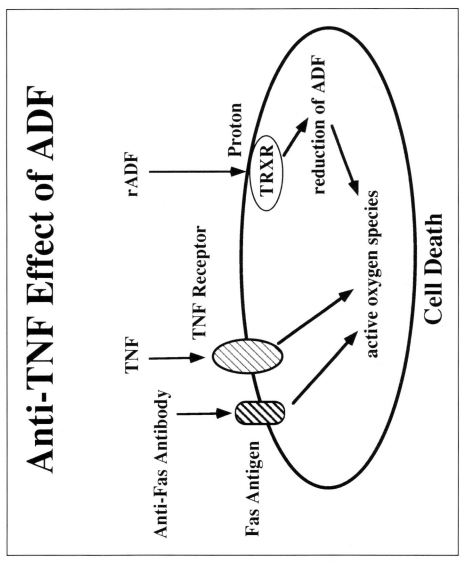

Fig. 7.3. Mechanism of cytoprotection by ADF against TNF and anti-Fas antibody.

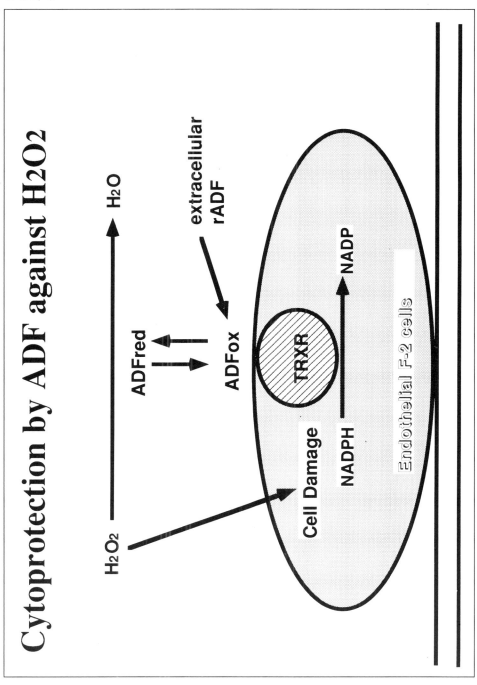

Fig. 7.4. Mechanism of cyto-protection by ADF against hydrogen peroxide.

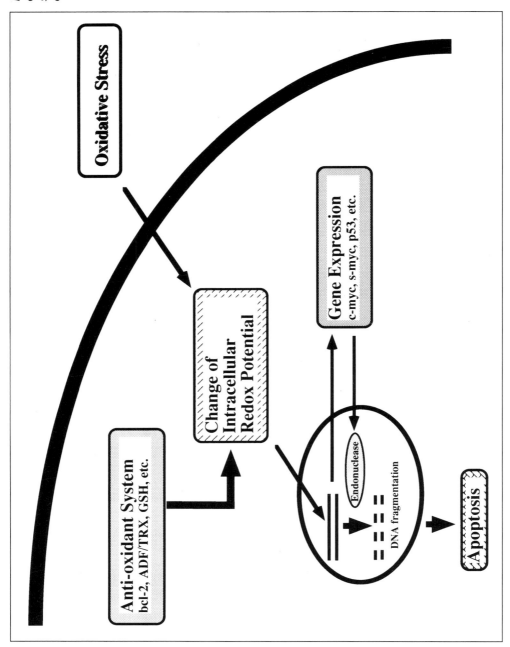

Fig. 7.5. Effect of oxidative stresses on the cells.

DEPENDENCY OF IMMUNE SYSTEM ON THE REDOX POTENTIAL

Cellular redox potential is maintained by several mechanisms including glutathione (GSH) and ADF/TRX systems.

It is well known that lymphoid cells are highly dependent on reduced conditions for their growth and activation. For example, without thiol compounds such as GSH and cystine, in vitro growth of lymphocytes is impossible, even in the presence of growth factors and mitogens as demonstrated by Yamauchi et al and Iwata et al.[9-11]

In the case of lymphocytes of various species including mouse and rabbit, thiols and disulfides such as 2-ME (mercaptoethanol), cysteine, and GSH have been shown to enhance their cytokine and/or mitogen-dependent proliferation by Fanger et al, Broome et al and Hewllet et al.[12-14] In addition, Hewllet et al suggested a possible role of macrophages in the generation of 2-ME-like factor.[14]

Proliferation and activation of human T cells by CD2 and CD3 antigens are dependent on glutathione contents and their reactivity could be sorted on the basis of glutathione contents.[15,16]

This thiol-dependency of immune responses has been known for years without much discussion. Recently, however, the importance of redox control of cell growth and cell death has been recognized in relation to retroviral disorders and programmed cell death or apoptosis.

In human T cells and T cell lines, thiols are required for the progression cell cycle. Iwata et al showed that transferrin receptor induction was not observed in thiol-free culture condition, while IL-2R α chain (Tac) expression is induced, implicating that T cell cycle is arrested at late G_1 phase. They reported that addition of rADF/TRX to the culture partially recovered this inhibition.[11]

Furthermore, ADF/TRX and GSH systems appear to functionally complement each other on the cellular level. Recently it was found that rADF/TRX greatly facilitate the internalization of cystine which is GSH precursor. In the presence of rADF/TRX, cystine uptake to the cells was rapidly augmented, whereas albumin or active site mutant of rADF that has no reducing activity, failed to do so. In association with this cystine uptake, the intracellular level of GSH was also significantly augmented by administration of rADF into the culture medium under the low concentration of cystine (around 50 μM) (Iwata S; personal communication).

Part of the cytoprotective effect and growth promoting effect of rADF/TRX therefore seems to be related to this replenishment of GSH system. This is quite important considering the recent information by Dröge et al and others about the redox dysregulation and loss of GSH in AIDS patients (Fig. 7.6).[17]

VARIETY OF SOLUBLE FACTORS RELATED TO ADF

During the study of ADF, it became apparent that there were several research groups independently studying the same factors using different assay systems.

A particularly important example was the study of IL-1-like factor(s) produced by EBV (Epstein Barr Virus)-transformed human B cell lines by Drs. Hiro and Naomi Wakasugi in the laboratory of Professor Thomas Tursz in Gustave-Roussy Institute in France. During their study of IL-1-like factor produced by the high producer cell line (3B6), they were impressed by the similarity

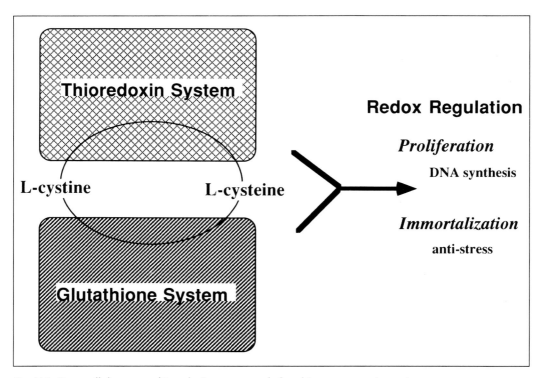

Fig. 7.6. Cross-talk between thioredoxin system and glutathione system via cystine transport.

of the properties between 3B6-IL-1 and ADF.[18] Wakasugi and Tursz contacted us before completion of the cDNA cloning for ADF and immediately started cooperation with us. It is now widely accepted that ADF and 3B6-IL-1 are the identical molecule human thioredoxin (hTRX). As is the case with HTLV-I+ T cells, rADF/TRX is also produced by EBV-transformed B cell lines and promotes the proliferation of these cell lines in vitro (Fig. 7.7).[19]

Independently, Silverstein and his colleague reported a cytokine named eosinophil cytotoxicity enhancing factor (ECEF), which later proved to be homologous to TRX.[20] ECEF was produced by activated macrophage cell line, U937, and augmented the cytotoxic activity of eosinophils. Hori et al in cooperation with Hirashima proved that rADF also enhanced the chemotactic activity of eosinophil cell lines.[21] It is interesting that ECEF activity of TRX was higher after a partial proteolysis. Such processing and modification may alter the biological activity and targeting properties of ADF/TRX (Table 7.1).

In cooperation with Holmgren, who initially described bacterial thioredoxins,[2] Rósen et al also reported that their B cell growth factor (BCGF) released from MP6 T hybridoma cell line seemed to be closely related to thioredoxin.[22,23] Thus, hTRX or ADF appear to have these different cytokine-like activities, depending on the assay systems used.

Table 7.1. Soluble factors related to human thioredoxin

ADF	ATL-derived factor	Teshigawara K et al	1985
3B6-IL-1	IL-1-like factor from 3B6 EBV+ B cell line	Wakasugi H et al	1987
ECEF	Eosinophil Cytotoxicity Enhancing Factor	Silverstein DS et al	1989
MP6-BCGF	BCGF from MP6 T cell hybridoma	Rósen et al	1986
EPF	Early Pregnancy Factor	Hegh V and Clunie GJA	1987
SASP	Surface Associated Sulfhydryl Protein	Martin H and Dean M	1991

Also interesting was the relationship of ADF to the serum factor(s) related to pregnancy. Hegh reported that Early Pregnancy Factor (EPF) inhibits T lymphocyte activity, however, the immunological significance of this remains unclear.[24] Tonisse and Wells recently reported that one of the components of the EPF complex is identical to hTRX.[25] As is discussed in the following paragraph, one of the tissues containing a high level of ADF/hTRX is female reproductive tissue including placenta.[26] As ADF/hTRX has an antioxystress activity, the reproductive system may require a reduced environment for ideal fertilization. Indeed, oxystress and reactive oxygen species (ROI) are known to inhibit in vitro fertilization.

REDOX DYSREGULATION IN ATL AND OTHER VIRAL INFECTIONS

ATL AND ADF; ENHANCED PRODUCTION OF ADF/TRX

ADF/TRX and IL-2R α chain (Tac) Overexpression In Vitro

Majorities of the HTLV-I-transformed T cell lines from ATL patients and HTLV-I-infected individuals are strongly expressing ADF. There seemed to be a rough correlation between morphological transformed phenotype and ADF production. Some IL-2-dependent HTLV-I⁺ T cell lines with normal lymphoid morphology are not strongly positive for ADF, while HTLV-I⁺ T lymphoblastoid cell types such as HUT102, ATL-2, and MT-2 are strongly positive for ADF.

ATL Derived Factor (ADF)
HTLV-1 induced T cell Transformation
Tagaya, Teshigawara, Yodoi et al.

3B6/IL-1
EBV-induced B cell Transformation
Wakasugi, Tursz et al.

Fig. 7.7. ATL-derived factor and 3B6-IL-1, proved to be identical protein by cDNA cloning.

The correlation between ADF production and the expression of IL-2R α chain and HTLV-I gene product such as p40^x/Tax is quite important. While majorities of ATL cell lines are strongly positive for ADF and IL-2R α chain (Tac), there are apparent exceptions. A typical example is MT-1 cells, which were T cell line cells infected with HTLV-I established by coculture with ATL cells and umbilical cord lymphocytes. The expression of ADF was not strongly augmented as compared with other HTLV-I-transformed T cell lines.

ADF overproduction may be one of the major mechanisms for the constitutive IL-2R α chain expression but an alternative mechanism may also be present in some situations. Also, not all the lymphocytes can express IL-2R α chain after ADF production. Indeed, in EBV-transformed B cells with ADF overproduction, FcεRII/CD23 is strongly induced whereas IL-2R α chain is induced only in part of the cases.[19,27]

On the cell line level, there may be a rough correlation between ADF and p40^x/Tax expression. However, on the single cell level, a precise analysis of the possible co-expression is needed for the clarification of the active mechanism in HTLV-I infection.

In Vivo Situation of ADF

There is no definite evidence for the constant up-regulation of ADF in ATL cells in vivo. However, in some cases, ADF-positive cells have accumulated in lymph nodes infiltrated with ATL cells. The identification of ADF-positive cells is important as ADF is induced in monocytes/macrophages and dendritic cells by immunological stimuli and stress. Both ATL cancer cells and reactive non-infected cells may be producing ADF in such pathological condition.

Further studies are needed to further eludicate the possible extracellular roles of ADF.

AIDS AND ADF; DEPLETION OF ADF/TRX AND OXYSTRESS IN HIV INFECTION

The Basic Study of Transport Systems for Thiols

Basic study of the transport of cysteine and cystine was carried out by Bannai et al.[28-31] Their analyses provide the rationale for the mechanism of how cystine/cysteine redox pairs regulate the functions of lymphocytes. According to them, lymphocytes have weak transport activity for cystine, which is called X_c system but strong transport activity for cysteine, which is called ASC system. In the plasma, however, the concentration of cysteine is maintained at low level. In contrast, activated macrophages (and epithelial cells) elicit strong transport activity for cystine and release cysteine into the extracellular space. Dröge et al recently showed that the cysteine supply from macrophages plays a regulatory role in thiol-dependent lymphocyte functions such as DNA synthesis, CTL activity in allogenic MLR, or LAK activity.[10,17] They also describe the inhibitory role of glutamate in Xc and subsequent cysteine release macrophages. Furthermore, the lymphoid system, neurons and glial cells have also been shown to interact through cysteine delivery. In fact, inhibition of cystine transport causes neuronal cell death (apoptosis) in the primary culture.

Cysteine and Glutathione Depletion in AIDS

In the past few years, decrease of intracellular glutathione (GSH) level in the T cells of HIV-infected individuals has been reported by several laboratories.[32,33]

Dröge and his coworkers first demonstrated that the cysteine supply to lymphocytes is impaired in some immunodificiency diseases such as AIDS and cancer.[34,35] They described reduced cysteine concentration and elevated glutamate concentration in the plasma samples from HIV-infected individuals. It was also reported that the intracellular concentration of GSH in PBMC and monocytes was reduced.

Dröge et al further demonstrated that extracellular administration of thiols (glutathione, cysteine, N-acetylcysteine) inhibited NF-κB activity in HIV-LTR and HIV-LTR-mediated transcriptional activation, by collaboration with Baeuerle who demonstrated the activation of NF-κB by H_2O_2 or ROI. On the basis of this observation, they suggest the possibility of clinical application of N-acetyl cysteine in AIDS and HIV-infected patients for replenishment of GSH depletion. N-acetyl cysteine (NAC) is the acetylated form of cysteine, which is easily internalized and utilized as the precursor of GSH by any cell type, in lieu of cystine and cysteine.[36-38]

Subsequently, their findings were confirmed by Herzenberg's laboratory.[39,40] They showed that CD4 and CD8 T cells with high intracellular glutathione levels were selectively lost according to the HIV infection progression, using flow cytometry for GSH. They also confirmed the N-acetyl cysteine (NAC)-mediated inhibition of NF-κB activation in HIV-LTR. A clinical trial of NAC has started under the control of Herzenberg's laboratory (personal communication).

The mechanism of this decrease is not yet fully clarified. However, it is known that extracellular redox status is important for the internalization of cysteine. As Bannai and Iwata suggested, there may be thiol-sensitive transporting systems on the cell membrane (Fig. 7.8).

Elimination of ADF High Producer Cells in HIV Infection

Histological study has shown that T cells in the lymphoid tissue express ADF/TRX in variable degrees depending on cell type. Whereas lymphocytes are poorly positive for ADF, macrophages and dendritic cells are ADF-positive. In inflammatory tissues, activated macrophages are strongly positive for ADF, while macrophages in normal lymphoid tissue are only weakly positive. In contrast, dendritic cells are strongly positive for ADF even in normal tissue. Particularly interesting is the human thymus, in which medullary dendritic cells produce high levels of ADF. Although we do not know the role of high concentrations of ADF in dendritic cell, it is possible that ADF may be involved in some unique function of dendritic cells. The possible role of ADF in the antigen presentation function has to be further studied.

In AIDS patients, ADF high producer cells were eliminated from the lymph nodes.[41] This is in accordance with the observation that dendritic cells were lost in the progression of the disorders. Dendritic cells have been implicated to be a reservoir of HIV virus. Therefore, it is very interesting to see whether ADF/TRX is important for some steps of immune responses. To see the effect of HIV infection on the production of ADF, we utilized two HTLV-I$^+$ T cell lines, SKT-1B and MT-2, both of which are ADF high producers which have been known to be infected by HIV. When HTLV-I$^+$ T cell line cells were superinfected with HIV, the production of ADF was markedly downregulated in 3 to 7 days after infection. The mechanism of ADF

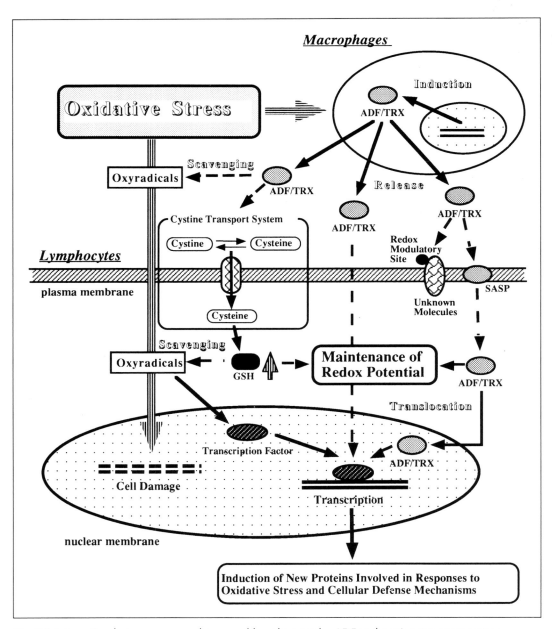

Fig. 7.8. Interaction between macrophages and lymphocytes by ADF and cystine transport system.

QUESTIONNAIRE

Receive a FREE BOOK of your choice

Please help us out—Just answer the questions below, then select the book of your choice from the list on the back and return this card.

R.G. Landes Company publishes five book series: *Medical Intelligence Unit, Molecular Biology Intelligence Unit, Neuroscience Intelligence Unit, Tissue Engineering Intelligence Unit* and *Biotechnology Intelligence Unit*. We also publish comprehensive, shorter than book-length reports on well-circumscribed topics in molecular biology and medicine. The authors of our books and reports are acknowledged leaders in their fields and the topics are unique. Almost without exception, there are no other comprehensive publications on these topics.

Our goal is to publish material in important and rapidly changing areas of bioscience for sophisticated scientists. To achieve this goal, we have accelerated our publishing program to conform to the fast pace in which information grows in bioscience. Most of our books and reports are published within 90 to 120 days of receipt of the manuscript.

Please circle your response to the questions below.

1. We would like to sell our *books* to scientists and students at a deep discount. But we can only do this as part of a prepaid subscription program. The retail price range for our books is $59-$99. Would you pay $196 to select four *books* per year from any of our Intelligence Units–$49 per book–as part of a prepaid program?

                   Yes   No

2. We would like to sell our *reports* to scientists and students at a deep discount. But we can only do this as part of a prepaid subscription program. The retail price range for our reports is $39-$59. Would you pay $145 to select five *reports* per year–$29 per report–as part of a prepaid program?

                   Yes   No

3. Would you pay $39–the retail price range of our books is $59-$99–to receive any single book in our Intelligence Units if it is spiral bound, but in every other way identical to the more expensive hardcover version?

                   Yes   No

To receive your free book, please fill out the shipping information below, select your free book choice from the list on the back of this survey and mail this card to:
  R.G. Landes Company, 909 S. Pine Street, Georgetown, Texas 78626 U.S.A.

Your Name _____

Address _____

City _____ State/Province: _____

Country: _____ Postal Code: _____

My computer type is Macintosh _____ ; IBM-compatible _____ ; Other _____

Do you own ____ or plan to purchase ____ a CD-ROM drive?

Available Free Titles

*Please check three titles in order of preference.
Your request will be filled based on availability. Thank you.*

- ☐ Water Channels
  Alan Verkman,
  University of California-San Francisco

- ☐ The Na,K-ATPase:
  Structure-Function Relationship
  J.-D. Horisberger, University of Lausanne

- ☐ Intrathymic Development of T Cells
  J. Nikolic-Zugic,
  Memorial Sloan-Kettering Cancer Center

- ☐ Cyclic GMP
  Thomas Lincoln, University of Alabama

- ☐ Primordial VRM System and the Evolution
  of Vertebrate Immunity
  John Stewart, Institut Pasteur-Paris

- ☐ Thyroid Hormone Regulation
  of Gene Expression
  Graham R. Williams, University of Birmingham

- ☐ Mechanisms of Immunological Self Tolerance
  Guido Kroemer, CNRS Génétique Moléculaire et
  Biologie du Développement-Villejuif

- ☐ The Costimulatory Pathway
  for T Cell Responses
  Yang Liu, New York University

- ☐ Molecular Genetics of Drosophila Oogenesis
  Paul F. Lasko, McGill University

- ☐ Mechanism of Steroid Hormone Regulation
  of Gene Transcription
  M.-J. Tsai & Bert W. O'Malley, Baylor University

- ☐ Liver Gene Expression
  François Tronche & Moshe Yaniv,
  Institut Pasteur-Paris

- ☐ RNA Polymerase III Transcription
  R.J. White, University of Cambridge

- ☐ src Family of Tyrosine Kinases in Leukocytes
  Tomas Mustelin, La Jolla Institute

- ☐ MHC Antigens and NK Cells
  Rafael Solana & Jose Peña,
  University of Córdoba

- ☐ Kinetic Modeling of Gene Expression
  James L. Hargrove, University of Georgia

- ☐ PCR and the Analysis of the T Cell Receptor
  Repertoire
  Jorge Oksenberg, Michael Panzara & Lawrence
  Steinman, Stanford University

- ☐ Myointimal Hyperplasia
  Philip Dobrin, Loyola University

- ☐ Transgenic Mice as an In Vivo Model
  of Self-Reactivity
  David Ferrick & Lisa DiMolfetto-Landon,
  University of California-Davis and Pamela Ohashi,
  Ontario Cancer Institute

- ☐ Cytogenetics of Bone and Soft Tissue Tumors
  Avery A. Sandberg, Genetrix & Julia A. Bridge,
  University of Nebraska

- ☐ The Th1-Th2 Paradigm and Transplantation
  Robin Lowry, Emory University

- ☐ Phagocyte Production and Function Following
  Thermal Injury
  Verlyn Peterson & Daniel R. Ambruso,
  University of Colorado

- ☐ Human T Lymphocyte Activation Deficiencies
  José Regueiro, Carlos Rodríguez-Gallego
  and Antonio Arnaiz-Villena,
  Hospital 12 de Octubre-Madrid

- ☐ Monoclonal Antibody in Detection and
  Treatment of Colon Cancer
  Edward W. Martin, Jr., Ohio State University

- ☐ Enteric Physiology of the Transplanted Intestine
  Michael Sarr & Nadey S. Hakim, Mayo Clinic

- ☐ Artificial Chordae in Mitral Valve Surgery
  Claudio Zussa, S. Maria dei Battuti Hospital-Treviso

- ☐ Injury and Tumor Implantation
  Satya Murthy & Edward Scanlon,
  Northwestern University

- ☐ Support of the Acutely Failing Liver
  A.A. Demetriou, Cedars-Sinai

- ☐ Reactive Metabolites of Oxygen and Nitrogen
  in Biology and Medicine
  Matthew Grisham, Louisiana State-Shreveport

- ☐ Biology of Lung Cancer
  Adi Gazdar & Paul Carbone,
  Southwestern Medical Center

- ☐ Quantitative Measurement
  of Venous Incompetence
  Paul S. van Bemmelen, Southern Illinois University
  and John J. Bergan, Scripps Memorial Hospital

- ☐ Adhesion Molecules in Organ Transplants
  Gustav Steinhoff, University of Kiel

- ☐ Purging in Bone Marrow Transplantation
  Subhash C. Gulati,
  Memorial Sloan-Kettering Cancer Center

- ☐ Trauma 2000: Strategies for the New Millennium
  David J. Dries & Richard L. Gamelli,
  Loyola University

downregulation is not clear at this moment, however, the differential effect of HTLV-I and HIV on the expression of ADF/TRX gene may be important.

REDOX CONTROL OF SIGNAL AND GENE EXPRESSION

Stress Inducibility and Function of ADF/TRX

As ADF was strongly expressed in HTLV-I and EBV transformed T and B cells, we first predicted that ADF is an inducible protein reactive to a variety of immunological as well as cell biological signals.

Upon lymphocyte activation by mitogens and PMA, other stimuli such as UV (ultraviolet) radiation, hydrogen peroxide (H_2O_2), and heavy metals such as cadmium have been shown to induce or enhance the expression of ADF.

Keratinocytes, endothelial cells, macrophages and lymphocyte cell lines are sensitive to oxidative stress such as UV, H_2O_2 and diamide. The expression of ADF/TRX protein and mRNA are strongly induced by these agents. The meaning of this oxystress-dependent induction is of interest, and implicates possible protective roles of this protein against oxystress and cytoprotection. Indeed, oxystress-dependent induction of ADF/TRX was confirmed in in vivo studies of rat retina and murine brain tissue. Particularly interesting is the induction of ADF/TRX in the selected cell types such as astroglia and pigment epithelium of respective tissues.[42,43]

The constitutive expression of ADF/TRX in dendritic cells of human thymic medulla as well as lymph nodes is interesting (Go et al unpublished). A possible role of ADF/TRX in antigen presenting function has been suggested in the study using ADF transgenic mice by Hori et al (in preparation).

Stress Inducible Element in ADF/TRX Promoter

Human ADF/TRX gene contains a unique oxystress-responsible element. Taniguchi have shown a 38 bp element reactive to H_2O_2, UV radiation, cadmium and other oxystress agents. The element appears to have no significant homology to AP-1 and other known elements. Unexpectedly, this element has no homology to heat shock element (HSE) (unpublished data).

Whether the oxystress-responsive element is involved in ADF

overexpression in the transformed cell infected with HTLV-I and EBV is an important question. Indeed, we have an unexpected evidence suggesting the alternative mechanism of ADF/TRX gene activation by viral gene products.

ADF/TRX as a Redox Regulator of Protein/Nucleotide Interaction

ADF/TRX has been shown to interact with various nucleotide binding proteins. First, Klausner et al showed that reducing condition by TRX is important for the binding of IRE (iron responsive element) with transferrin receptor mRNA.[44] Curran et al found that the binding of Jun/Fos complex to the AP-1 site of DNA is highly dependent on the reducing environment.[45] They showed that TRX greatly enhanced the in vitro binding. Xanthoudakis and Curran further demonstrated that Ref-1 gene product, a nuclear protein plays a similar role as TRX, which implicates a possible cascade of redox regulation involving endogenous TRX and thiol-active nuclear proteins.[46] Direct cysteine-mediated interaction between Ref-1 and Jun has recently been.[47] Quite interestingly, Ref-1 protein belongs to a group of DNA repair proteins named class II hydrolytic apurinic/apyrimidinic (A/P) endonucleases and has a potent DNA repair activity.[48,49] The involvement of redox regulation in the DNA repair mechanism is one of the major topics of the related field (Fig. 7.9).

Before the cloning of ADF/TRX genes, biochemical studies have strongly indicated that DNA-binding of glucocorticoid receptor/hormone complex required additional endogenous protein such as TRX. Therefore, it is a tempting notion that redox regulation involving ADF/TRX is a general mechanism required for the effective interaction between proteins and nucleotides.

NF-κB Dysregulation in HIV Infection

There is a reduction of T helper cells (T4) in PBL of HIV-infected individuals. While increase of TNF-α is present in these patients, however partly due to various infections.

Baeuerle et al first reported that NF-κB activation involved an oxidative stress of the cells.[50] In a subclone of Jurkat T cells, they demonstrated that H_2O_2 activated the intracellular NF-κB.

Similar results were confirmed by Roederer in Herzenberg's laboratory.[40] Although the critical steps of NF-κB activation by

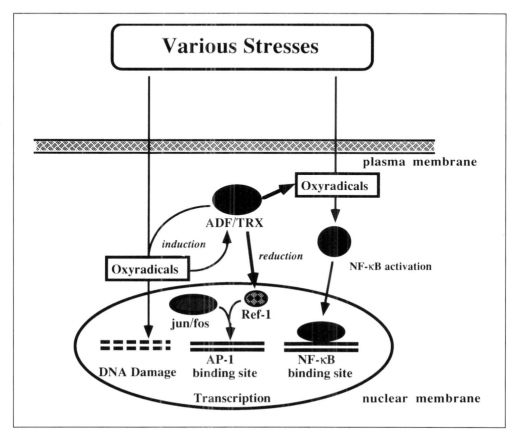

Fig. 7.9. Oxidative stress-mediated gene activation and ADF.

H_2O_2 are still unclear, continued activation of NF-κB in the lymphocytes of HIV-infected patients by cytokinemia and overwhelming antigenic challenges were hypothesized to be the cause of T4 cell elimination and immune dysfunction. In cooperation with us, Okamoto showed that in vitro interaction between NF-κB binding site of HIV LTR and NF-κB protein is strongly augmented by ADF/TRX.[51]

As mentioned above, Dröge et al and Baeuerle also showed that extracellular administration of thiols (glutathione, cysteine, N-acetyl cysteine) inhibited NF-κB activity in HIV LTR and it mediated transcriptional activation.

Therefore it is highly likely that redox dysregulation of NF-κB is deeply involved in the progression and pathophysiology of AIDS. The importance of thiols including GSH, cysteine and ADF/TRX in both intra- and extracellular compartments is apparent. Schenk et al recently showed the differential roles of these thiol compounds including thioredoxin in different compartments, indicating the necessity of intra- and extracellular redox controlling mechanisms.[51a]

INTRACELLULAR TRANSLOCATION OF ADF/TRX

Because of the relative ignorance about intranuclear localization of TRX in early studies, the idea of how ADF/TRX can interact with DNA was puzzling. Despite the ample evidence of redox regulation of protein/nucleotide interaction by ADF/TRX and other thiols, as demonstrated in AP-1, NF-κB and glucocorticoid receptors, there has been criticism that such regulation may not be important in vivo partly due to a lack of information about the intracellular redox environment.

From the standpoint of intracellular translocation of proteins, there is a unique property of ADF/TRX, which shuttles between cytosolic and nuclear compartments. In the histological analysis of precancerous tissue of female cervical mucosa, Fujii et al showed that the topology of ADF protein was variable; ADF in the epithelial cells in the basal layer was mainly in the nucleus, whereas ADF in the differentiated cells in the superficial layer was in the cytosol.[52] Their observation naturally indicated the importance of intracellular localization of ADF/TRX.

Recently, strong evidence for the translocation of the ADF/TRX has been obtained in in vitro culture of keratinocytes. Wakita et al observed that ultraviolet B (UVB) radiation induced the rapid translocation of cytosolic ADF/TRX into the nuclear compartment. This translocation was followed by slower induction of ADF/TRX production (unpublished). This sequential change of intracellular localization of ADF/TRX and neogenesis is in coherence with the process of transcription factors such as NF-κB and related rel gene products, and indicates the possible involvement of redox regulation with the translocation mechanism. A preliminary study with HTLV-I+ T cell lines disclosed quite recently that there may be a cell cycle-dependent translocation of ADF/TRX (Ueda-Taniguchi Y et al, in preparation).

Regulation of ADF/TRX gene expression as well as the redox state of ADF/TRX protein may be a new possible direction to correct deregulated cell activation and growth. This is dicussed further in the section on the pharmacological study of ADF/TRX regulation.

OXYSTRESS AND MEMBRANE SIGNALING; CROSS TALK WITH PHOSPHORYLATION

The catalogue of the target molecules of redox regulation is expanding. Oxystress by H_2O_2 and other agents such as diamide has been shown to directly trigger membrane tyrosine kinases. Nakamura et al in our laboratory showed that in Jurkat human T cell line cells as well as peripheral blood T cells, oxystress rapidly phosphorylates tyrosine kinase Lck.[53] Concerning the biological function, oxystress appears to be involved in programmed cell death, apoptosis, of the cells. In some cases, however, oxystress by $HgCl_2$ gives a proliferating signal to the cells as shown by Nakashima et al.[54] The mechanism of this activation is still

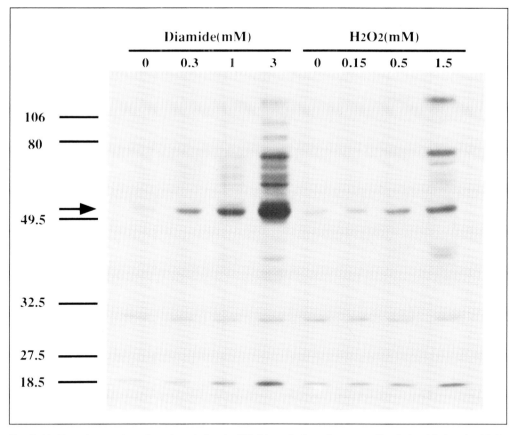

Fig. 7.10. Protein-tyrosine phosphorylation in PBL blasts induced upon sulfhydryl oxidation by H_2O_2 and diamide. Comparison of tyrosine phosphorylation induced by diamide and H_2O_2. Cells were treated with various doses of H_2O_2 for 2 minutes or diamide for 10 minutes at 37°C and 20 µg of whole cell lysates were loaded in each lane. Gel electrophoresis and immunoblotting with anti-phosphotyrosine antibody were performed. The arrow indicates the 55 kDa molecule.

unclear, although oxystress-induced inactivation of related phosphatases may be involved. This phenomenon strongly indicates that redox regulation and phosphorylation/dephosphorylation of signaling molecules are tightly associated (Figs. 7.10, 7.11).

The possible involvement of redox active thiols such as GSH and ADF/TRX in regulation of the oxystress-dependent signaling process needs to be clarified. This thiol-specific antioxidant may have a role in protecting the cells from extracellular oxystress. Indeed, ADF/TRX has been known to be concentrated in the cell membrane.

Redox regulation is therefore operating mainly on two distinct steps of signal transduction; one on the membrane signaling involving tyrosine kinases and the other on the level of protein/nucleotide interaction involving transcriptional factors. It is important to note that dysregulation of the redox environment in pathological situations such as inflammation and infection may profoundly affect immune responses.

VIRUS-RELATED TRANSFORMATION

T Lymphocytes and HTLV-I

As discussed fully in this chapter, the relationship between ADF/TRX and HTLV-I infection is apparent, although the molecular mechanism is not yet fully clarified.

B Lymphocytes with Epstein-Barr Virus (EBV)

Enhanced production of ADF/TRX was first clarified by Wakasugi and Tursz, who had reported as IL-1-like factor from EBV-transformed B cell lines.[18] In cooperation with us, their B cell-derived factor called 3B6-IL-1 subsequently proved to be identical to ADF.[6]

In an in vivo study using NPC (nasopharyngeal carcinoma), constitutive production of TRX/ADF in the transplanted tumor was demonstrated in nude mice. rADF/TRX and even bacterial TRX enhanced the growth of EBV+ B serum-deficient cell lines (1% serum). Under such conditions, Wakasugi demonstrated that proliferative response to IL-1 and to other cytokines was markedly enhanced by TRX/ADF.[19]

IL-2R α chain and FcεRII/CD23 expression on these cells were also enhanced in some EBV-transformed B cell lines. Indeed, it is

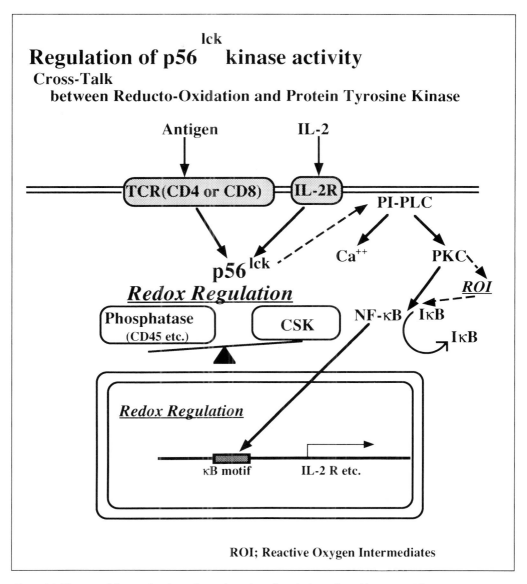

Fig. 7.11. The possible mechanism of tyrosine phosphorylation of p56^{lck} upon oxidative stress.

well known that FcεRII/CD23 is an activation antigen of B cells and is strongly induced by EBV infection as well as allergic condition involving IL-4 and IgE. ADF/TRX may be required for the activation of EBV-infected B cells. Comparative study of EBV-related and HTLV-I-related oncogenesis of B and T cells is a quite informative strategy for the elucidation of both disorders (Fig. 7.12).

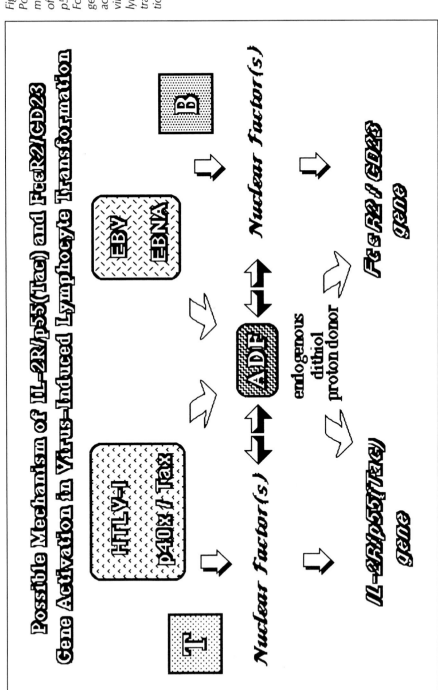

Fig. 7.12. Possible mechanism of IL-2R/p55(Tac) and FcεRII/CD23 gene activation in virus-induced lymphocyte transformation.

Epithelial Tissue and Papilloma

In malignant transformation of epithelial cells, expression of ADF/TRX is increased, as first demonstrated by Fujii et al in cervical mucosal epithelium.[52] Kusama et al showed that ADF/TRX was overexpressed in dysplasia and in an early stage of cancer of the oral cavity.[55] Similar overexpression of ADF/TRX is observed in dysplasia and cancer of the skin as shown by Wakita et al.[56] Consensus is that ADF/TRX expression is strongly enhanced in the early stage of malignant transformation of epithelium. Particularly interesting was the finding of Fujii et al that co-localization of HPV (human papilloma virus) DNA expression and ADF/TRX in precancerous tissue and dystrophy of female cervical mucosa. In the cells with cytopathic changes, both HPV DNA and ADF/TRX was strongly expressed. Although the relationship of this co-expression has yet to be clarified, ADF/TRX may be protective against the pathological processes and stress. The possible clinical utility of ADF/TRX as a marker of epithelial dystrophy and precancerous changes is suggested by these findings.

Liver and Hepatoma

The expression ADF/TRX is also significant in liver. Nakamura showed that ADF/TRX is strongly expressed in transformed cell lines of hepatocytes.[57] The direct relationship between hepatoma virus and ADF/TRX is unclear, however. Regeneration of liver after partial hepatectomy of rat is associated with ADF/TRX induction (Inamoto et al personal communication). In in vitro maintenance of normal rat hepatocytes, there is a significant induction of ADF/TRX, indicating that stress conditions in culture environments may induce ADF/TRX gene expression. This is similar to the case with macrophage/monocytes, that express significant amount of ADF/TRX in culture.

There is a close relationship between ADF/TRX expression and transformation and dysplasia of variety of cell types. However, mechanism and biological meaning of this ADF/TRX induction may vary depending on the cell types and viruses. Although ADF/TRX may provide protection against cytopathic effects and oxystress generated by viral gene products, there is no direct evidence for this.

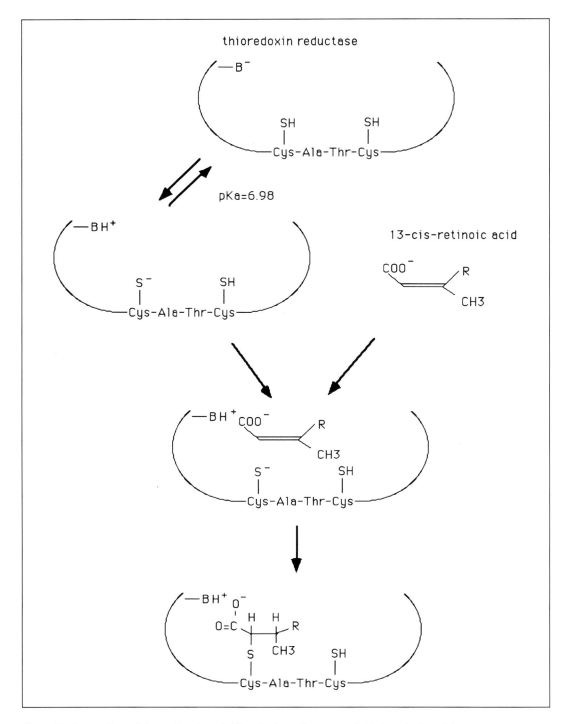

Fig. 7.13. Interaction of the active site of thioredoxin reductase and 13-cis-retinoic acid.

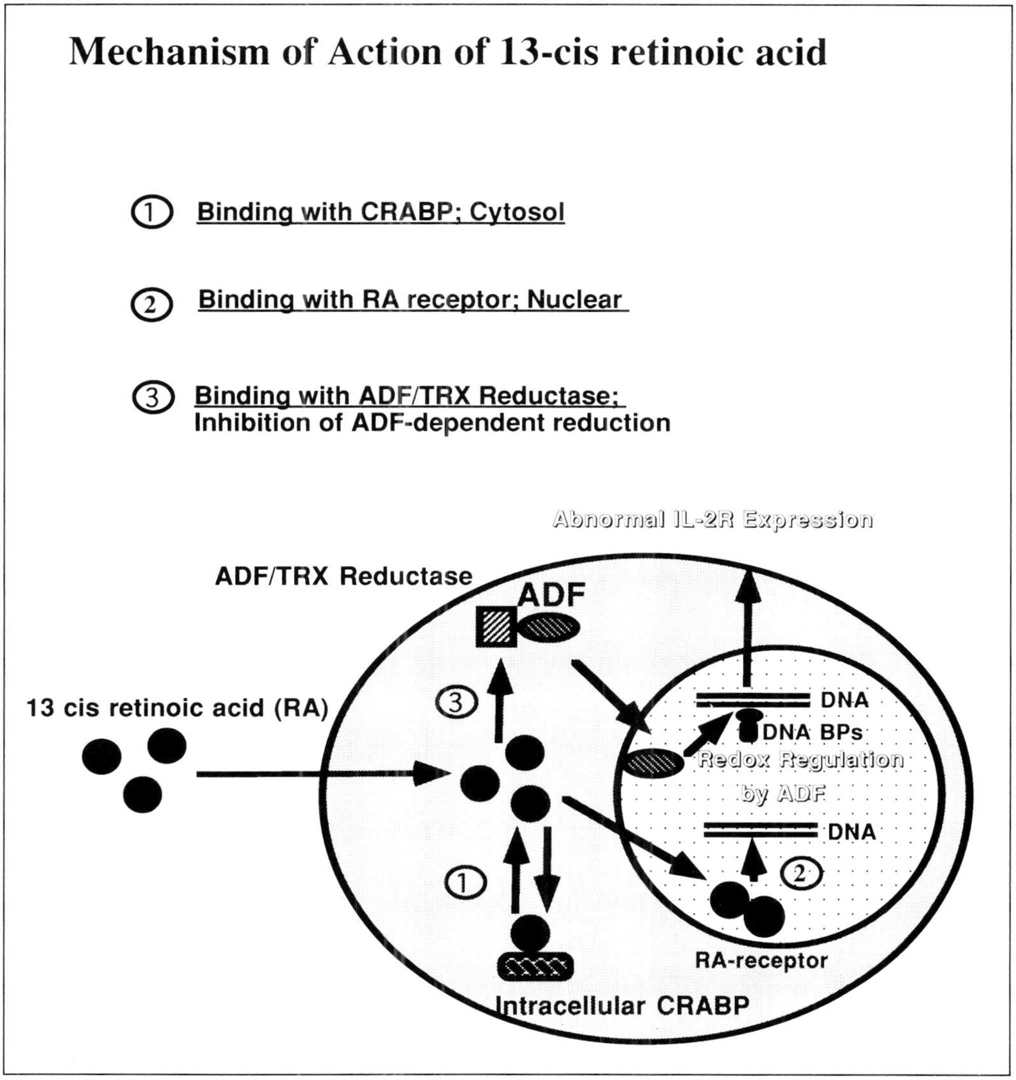

Fig. 7.14. Mechanism of action of 13-cis-retinoic acid on the cells.

SERUM ADF/TRX

ADF was originally found as an extracelluar cytokine-like factor. Other investigators also independently reported TRX as a soluble mediator, as 3B6-IL-1 from B cells, MP6-BCGF from T hybridoma cells, ECEF from macrophage-like cell line cells, and EPF detected in the serum of pregnant women.

Therefore, it is quite important as to how the intracellular TRX can be released from the cells into extracellular compartments both in vivo and in vitro. Despite the lack of signal sequence for secretion, there is evidence strongly indicating the active process to release ADF/TRX mRNA. Martin et al reported the TRX-related protein tightly associated with plasma membrane as SASP (surface associated sulfhydryl protein).[58] SASP was easily released into the culture medium upon mild treatment of the cells such as a pH shift. In the study of the protection of neurons against ROI and ischemia, Hori recently found that neuronal cells actively release ADF/TRX when exposed to oxystress by H_2O_2.[59] It appears that oxidative stresses not only induce the production of ADF/TRX but also activate the release of ADF/TRX.

Kitaoka formerly demonstrated the presence of ADF/TRX in the serum using ELISA.[60] In a variety of disorders including ATL and hepatic diseases and some autoimmune disorders, there was an increase of serum ADF/TRX. The correlation of serum ADF/TRX to the clinical state and the origin of serum ADF/TRX is under investigation.

The in vivo biological significance of extracellular ADF/TRX is still unknown. However, the protective role of this protein against oxidative stress and cytotoxic molecules has been generated in local microenvironment. Quite interestingly, in a study of soluble factors active on a cholinergic neuron, Endoh in Tabira's group recently found that one of the active principles with trophic activity was TRX.[61] The anti-oxystress effect may explain some of the immune potentiating activity and anti-apoptotic activity of ADF/TRX, although further study is needed to determine the actual biological function of this protein in extracellular compartments.

CLINICAL UTILITY OF REDOX CONTROL AND ADF/TRX

Because of the variety of the biological activities of ADF/TRX, which has a relatively low molecular size of 13 kDa, one immediate possibility is to utilize ADF/TRX as a unique redox regulatory protein. Indeed, Wada et al has shown that ADF has a potent protective activity against reperfusion injury of lung in rabbit and dog.[62]

ADF/TRX expression in activated T cells is inhibited by pharmacological agents such as cyclosporin A (CysA) and FK506 (Furuke K et al, submitted). In the presence of CysA and FK506, mitogen-dependent induction of ADF/TRX in PBLs was inhibited. However, ongoing production of ADF/TRX by HTLV-I transformed T cells was not inhibited as was the case with activated T cells. The immunosuppressive effects of CysA and FK506 were also partially inhibited by ADF in culture, implicating the involvement of redox regulation in the action of these agents.

Retinoids also influence retinoid derivatives that are inhibitors of thioredoxin reductase. HTLV-I$^+$ T cell line cells highly producing ADF/TRX were sensitive to retinoids as compared with HTLV-I$^-$ T cell lines. Proliferation of PBL in the patients with HAM and leukemic cells from the patients with ATL was inhibited by retinoids, although the mechanism has yet to be clarified. The treatment of ATL and HTLV-I-related disorders using retinoid derivative is under investigation (Figs. 7.13, 7.14).

REFERENCES

1. Tagaya Y, Maeda Y, Mitsui A et al. ATL-derived factor (ADF), an IL-2 receptor/Tac inducer homologous to thioredoxin; Possible involvement of dithiol-reduction in the IL-2 receptor induction. EMBO J 1989; 8:757-64.
2. Holmgren A. Thioredoxin. Ann Rev Biochem 1985; 54:237-71.
3. Tagaya Y, Masutani H, Nakamura H et al. Role of ATL-derived factor (ADF) in the normal and abnormal cellular activation. Involvement of dithiol related reduction. Mol. Immunol. 1990; 27:1279-89.
4. Mitsui A, Hirakawa T, Yodoi J. Radical scavenging activity and protein refolding activity of recombinant adult T cell leukemia-derived factor (ADF)/Human thioredoxin. Biochem Biophysi Res Comm 1992; 186:1220-1226.
5. Yodoi J, Uchiyama T. Human T-cell Leukemia Virus Type I (HTLV-I) associated diseases; virus, IL-2 receptor dysregulation and redox regulation. Immunology Today 1992; 13:13:405-411.
6. Yodoi J, Tursz T. ADF: An endogenous reducing protein homologous to thioredoxin; Involvement in lymphocyte immortalization by HTLV-I and EBV. Adv Cancer Res 1991; 57:381-411.
7. Mitsuhiro M, Masutani H, Nakamura H et al. Protective activity of ATL-derived factor (ADF) against TNF-dependent cytotoxicity on U937 cells. J Immunology 1991; 147:3837-3841.
8. Nakamura H, Matsuda M, Furuke F et al. Inhibition of activated neutrophil- or H^2O^2-induced cytotoxicity on murine endothelial F-2 cells by adult T cell leukemia-derived factor/human thioredoxin.

Immunol Letter (in press).
9. Yamauchi A, Nakamura Y, Inamoto T et al. Liver diseases and hyporesponsiveness to a new mitogenic monoclonal antibody YTA-1 recognizing p75 molecule on LGL/NK cells; Possible correlation with mitochondrial dysfunction. Molecular Immunol 1992; 29:263-70.
10. Yamauchi A, Bloom ET. Requirement of thiol compounds as reducing agents for IL-2 mediated induction of LAK activity and proliferation of human NK cells. J Immunol 1993; 151:5535-44.
11. Iwata S, Hori T, Sato N et al. Thiol-mediated redox regulation of lymphocyte proliferation. Possible involvement of adult T cell leukemia-derived factor and glutathione in transferrin receptor expression. J Immunol 1994; 152:5633-42.
12. Fanger MW, Hart DA, Wells JV et al. Enhancement by reducing agents of the transformation of human and rabbit peripheral lymphocytes. J Immunol 1970; 105:1043-5.
13. Broome JD, Jeng MW. Promotion of replication in lymphoid cells by specific thiols and disulfides in vitro. J Exp Med 1973; 138: 574-92.
14. Hewllet G, Opitz HG, Schlumberger HD et al. Growth regulation of a murine lymphoma cell line by a 2-mercaptoethanol or macrophage-activated serum factor. Eur J Immunol 1977; 7:781-5.
15. Kavanagh TJ, Grossmann A, Jaecks EP et al. Proliferative capacity of human peripheral blood lymphocytes sorted on the basis of glutathione content. J Cell Physiol 1990; 145:472-80.
16. Suthanthiran M, Anderson ME, Sharma Y et al. Glutathione regulates activation-dependent DNA synthesis in highly purified normal human T lymphocytes stimulated via the CD2 and CD3 antigens. Proc Natl Acad Sci USA 1990; 87:3343-7.
17. Dröge W, Eck HP, Gmünder H, Mihm S. Modulation of lymphocyte functions and immune responses by cysteine and cysteine derivatives. Am J Med 1991; 3C:140S-144S.
18. Wakasugi H, Rimsky L, Mahe Y et al. Epstein-Barr virus-containing B-cell line produce an interleukin 1 that it uses as a growth factor. Proc Natl Acad Sci USA 1987; 84:804-8.
19. Wakasugi N, Tagaya Y, Wakasugi H et al. Adult T-cell leukemia-derived factor/thioredoxin, produced by both human T-lymphotropic virus type I- and Epstein-Barr virus-transformed lymphocytes, acts as an autocrine growth factor and synergizes with interleukin 1 and interleukin 2. Proc Natl Acad Sci USA 1990; 87:8282-6.
20. Silverstein DS, Ali MH, Baker SL et al. Human eosinophil cytotoxicity-enhancing factor. Purification, physical characteristics, and partial amino acid sequence of an active polypeptide. J Immunol 1989; 143:979-83.
21. Hori K, Hirashima M, Yodoi J. Regulation of eosinophil migration by adult T cell leukemia-derived factor (ADF). J Immunol

1993; 151:5624-30.
22. Rosén A, Uggla C, Szigeti R et al. A T-helper cell X Molt4 human hybridoma constitutively producing B-cell stimulatory and inhibitory factor. Lymphokine Research 1986; 5:185-204.
23. Carlsson M, Sundstrom C, Bengtsson M et al. Interleukin 4 strongly augments or inhibits DNA synthesis and differentiation of B-type chronic lymphocytic leukemia cells depending on the co-stimulatory activation and progression signals. Eur J Immunol 1989; 19:913-21.
24. Morton H, Hegh V and Clunie GJA. Immunosuppression detected in pregnant mice by rosette inhibition test. Nature 1974; 249:459-60.
25. Tonisse KF, Wells JRE. Isolation and characterization of human thioredxin-encoding genes. Gene 1991; 102:221-8.
26. Fujii S, Yoshihiko N, Nonogaki H et al. Immunohistochemical localization of ATL-derived factor (ADF), a human thioredoxin homologue, in human fetal tissues and placentas. Virchows Archiv A, Pathological Anatomy and Histopathology 1991; 49:317-326.
27. Ikuta K, Takami M, Choong-won K et al. Cloning of cDNA for human lymphocyte Fce receptor; Homology with animal lectins. Proc Natl Acad Sci USA 1987; 84:819-23.
28. Bannai S, Kitamura E. Transport interaction of L-cystine and L-glutamate in human diploid fibroblasts in culture. J Biol Chem 1980; 255: 2372-6.
29. Bannai S. Transport of cystine and cysteine in mammalian cells. Biochimica et Biophysica Acta 1984; 79:289-306.
30. Miura K, Ishii T, Sugita Y et al. Cystine uptake and glutathione level in endothelial cells exposed to oxidative stress. Am J Physiol 1992; 262:C50-8.
31. Bannai S. Exchange of cystine and glutamate across plasma membrane of human fibroblasts. J Biol Chem 1986; 261:2256-63.
32. Buhl R, Jaffe HA, Holroyd KJ et al. Systemic glutathione deficiency in symptom-free HIV-seropositive individuals. Lancet 1989; ii:1294-1298.
33. Wu J, Levy EM, Black PH. 2-Mercaptoethanol and N-acetylcysteine enhance T cell colony formation in AIDS and ARC. Clin Exp Immunol 1989; 77:7-10.
34. Dröge W, Eck HP, Naher H et al. Abnormal amino acid concentration in the blood of patients with acquired immunodeficiency syndrome (AIDS) may contribute to the immunological defect. Biol Chem Hoppe-Seyler 1988; 369:143-148.
35. Eck HP, Gmünder H, Hartmann M et al. Low concentration of acid-soluble thiol (cysteine) in the blood plasma of HIV-1 infected patients. Biol Chem Hoppe-Seyler 1989; 370:101-108.
36. Dröge W, Eck HP, Mihm S. HIV-induced cysteine deficiency and T-cell dysfunction-a rationale for treatment with N-acetylcysteine.

Immunology Today 1992; 13: 211-214.
37. Mihm S, Ennen J, Pessara U et al. Inhibition of HIV-1 replication and NF-κB activity by cysteine and cysteine derivatives. AIDS 1991; 5:497-503.
38. Mihm S, Dröge W. Intracellular glutathione level control DNA-binding activity of NF-κB-like proteins. Immunobiology 1990; 181:245.
39. Staal FJT, Roederer M, Herzenberg LA et al. Intracellular thiols regulate activation of nuclear factor κB and transcription of human immunodeficiency virus. Proc Natl Acad Sci USA 1990; 87; 9943-7.
40. Roederer M, Staal FJT, Osada H et al. CD4 and CD8 T cells with high intracellular glutathione levels are selectively lost as the HIV infection progresses. Int Immunol 1991; 3:933-7.
41. Masutani H, Naito M, Takahashi K et al. Dysregulation of adult T cell leukemia-derived factor (ADF)/thioredoxin in HIV infection: Loss of ADF high producer cells in lymphoid tissues of AIDS patients. AIDS Research and Human Retroviruses 1992; 8:1707-15.
42. Tomimoto H, Akiguchi I, Wakita H et al. Astroglial expression of ATL-derived factor, a human thioredoxin homologue, in the gerbil after transient global ischemia. Brain Research 1993; 625:1-8.
43. Ohira A, Honda O, Chiaki D Gauntt et al. Oxidative stress induces adult T leukemia-derived factor (ADF)/Thioredoxin in the rat retina. Laboratory Investigation 1994; 70:279-85.
44. Hentze MW, Rouault TA, Harford JB et al. Oxidation-reduction and molecular mechanism of a regulatory RNA-protein interaction. Science 1989; 244:357-9.
45. Abate CL, Patel FJ, Rauscher III et al. Redox regulation of Fos and Jun DNA-binding activity in vitro. Science 1990; 249:1157-61.
46. Xanthoudakis S, Curran T. Identification and characterization of Ref-1, a nuclear protein that facilitates AP-1 DNA-binding activity. EMBO J 1992; 11:653-65.
47. Xanthoudakis S, Miao GG, Curran T. The redox and DNA-repair activities of Ref-1 are encoded by nonoverlapping domains. Proc Natl Acad Sci USA 1994; 91:23-7.
48. Demple B, Herman T, Chen DS. Cloning and expression of APE, the cDNA encoding the major human apurinic endonuclease: Definition of a family of DNA repair enzymes. Proc Natl Acad Sci USA 1991; 88:11450-54.
49. Seki S, Hatsushika M, Watanabe S et al. cDNA cloning, sequencing, expression and possible domain structure of human APEX nuclease homologous to Echerichia coli exonuclease III. Biochem et Biophysica Acta 1992; 1131:287-99.
50. Schreck R, Rieber P, Baeuerle PA. Reactive oxygen intermediates as apparently widely used messengers in the activation of the NF-κB transcription factor and HIV-1. EMBO J 1991; 10:2247-58.
51. Okamoto T, Ogiwara H, Hayashi T et al. Human thioredoxin/

adult T cell leukemia-derived factor activates the enhancer binding protein of human immunodeficiency virus type 1 by thiol redox control mechanism. International Immunology 1992; 4:811-9.

51a. Schenk H, Klein M, Erdbrügger W et al. Distinct effects of thioredoxin and antioxidants on the activation of transcription factors NF-κB and AP-1. 1994; 91:1672-6.

52. Fujii S, Yoshihiko N, Nonogaki H. et al. Co-expression of ATL-derived factor (ADF), a human thioredoxin homologue, and human papilloma virus (HPV) DNA in neoplastic cervical squamous epithelium. Cancer 1991; 68:1583-91.

53. Nakamura K, Hori T, Sato N et al. Redox regulation of p56lck protein tyrosin kinase activity in T cells; Involvement of sulfhydryl oxidation in the activation of src family protein kinase p56lck in T cells. Oncogene 1993; 8:3133-9.

54. Nakashima I, Pu M, Nishizaki A et al. A redox mechanism as alternatve to ligand binding for receptor activation delivering disregulated cellular signals. J Immunol 1994; 152:1064-71.

55. Kusama K, Saitoh T, Masutani H et al. Adult T cell leukemia derived factor (ADF) in oral epithelial lesions. J Oral Pathol Med 1991; 20:421-4.

56. Wakita H, Yodoi J, Masutani H et al. Immunohistochemical distribution of adult T-cell leukemia derived factor/thioredoxin in epithelial components of normal and pathologic human skin conditions. J Investigative Dermatology, 1992; 99:101-7.

57. Nakamura H, Masutani H, Tagaya Y et al. Expression and growth promoting effect of adult T cell leukemia derived factor (ADF), a human thioredoxin homologue in hepatocellular carcinoma. Cancer 1992; 69:2091-7.

58. Martin H and Dean M. Identification of a thioredoxin-related protein associated with plasma membrane. Biochem Biophysi Res Comm. 1991; 175:123-7.

59. Hori K, Katayama M, Sato N et al. Neuroprotection by glial cells through ATL-derived factor/human thioredoxin (ADF/TRX). Brain Research (in press).

60. Kitaoka Y, Sorachi K, Nakamura H et al. Detection of adult T cell leukemia-derived factor (ADF)/human thioredoxin in human serum. Immunology Letter (in press).

61. Endoh M, Kunishita T, and Tabira T. Thioredoxin from activated macrophages as a trophic factor for central cholinergic neurons in vitro. Biochem Biophysi Res Comm 1993; 192:760-5.

62. Yokomise H, Fukuse T, Hirata T et al. Effect of recombinant human Adult T-Cell Leukemia-Derived Factor (ADF) on rat lung reperfusion injury. Respiration 1994; 61:99-104.

INDEX

Page numbers in italics denote figures (f).

A

Acquired immunodeficiency syndrome (AIDS), 1, 3, 12, 14, 39, 108, 112, 117. *See also HIV*.
  ADF and, 111-115
  comparison with ATL-HTLV-I relationship, 1
  cysteine and glutathione depletion and, 112-113
Adult T cell leukemia (ATL)
  abnormal activation phenotype, 2
  acute type (acute ATL), 2, 17-18, 21, 24-25, 27, 48, 73, 87
  adult T cell leukemia (ATL), 42, 126
  ATL-2, 110
  ATL/SCID model, 95-97
  ATL/SCID studies, summary of, 94
  -associated antigen (ATLA), 9
  carcinogenesis model, multi-step, 82
  cell(s)
    ADF, upregulation of, in, 111
    augmented production of cytokines in, 2
    biological properties of, 45-54
      surface phenotype and function, in vitro, 45-50
      cytokines and other peptides produced by, 50-54
    CD3 complex, expression of, 48-49
    immune function of, in vitro, 53
    lines from, 6, 9
    morphological features of, 6
    peptides produced by, 50
    proliferation
      animal model of, 87
      in vivo, 81-97
      radiometric studies on, 95-97
    suppressive activity of, 53
    T cell antigen receptor, expression of in, 48
  chronic type (chronic ATL), 2, 17-18, 20, 24, 27, 48, 73
  classification, 17-21
  clinical features of, 2, 17-32
    laboratory findings, 22-26
    symptoms, 21
  complications, 26-27
  diagnostic criteria for clinical subtypes, 18, 20-21, 73
  etiology, 17
  history of, 5-14
  HTLV-I, relationship with, 81, 95
  incidence, 13, 17, 81
  infections, 26
  intermediate groups, 2
  lymphoma type, 17, 19, 21, 24-25, 27, 73, 90
  mechanism of oncogenesis in, 81
  pathology of, 6
  patients
    geographical distribution of, in Japan, 7
    histopathological findings of a lung of, 22-23
    HTLV-I cell lines, from ATL, 10
    leukemic cells from, 24, 45-46, 49-51, 53, 67-69, 71, 84-91, 127
    osteolytic lesions of the skull of, 25
    sera from, 9, 31
  prognosis, 26, 27, 87
  redox dysregulation in, 12, 110-114
  skin, leukemic cell infiltration into, 19, 21, 96
  smoldering type, 2, 17-18, 20, 24, 27, 73
  stress responses, in, 12
  studies
    Japan, in, 5-6
    U.S., in, 6-9
  subacute, 2
  treatment, 27-28, 50, 87, 127
    remission rates with, 28
Adult T cell leukemia (ATL)-derived factor (ADF). *See also HTLV-I*.
  ADF, 2, 3, 54

ADF/thioredoxin (TRX)
  ADF/TRX, 12, 42, 101-127
    cytoprotection, mechanism of, 101-103
    immune system dependency on, redox potential, 107-108
    insulin-reducing activity, 102
    interleukin-2 (IL-2)Rα expression in, 110-111
    intracellular translocation of, 118
    promoter, stress-inducible element, 115-116
    protein, expression of, 115
    redox control and clinical utility, 126-127
    redox regulation
      protein/nucleotide interactions, of, 116
      ADF, and, 101-127
    serum, 125-126
    stress-inducibility and function of, 115
  cysteine transport system, and, *114*
  downregulation, 113, 115
  high producer cells, 113-115
  hydrogen peroxide (H2O2), cytoprotection against, *105*
  interleukin-2 (IL-2)Rα
    expression, and enhancement of, 12
    HTLV-I gene product, and, 111
  in vivo, 111
  oxidative stress-mediated gene activation and, *117*
  overproduction of, 111, 123
  production, regulation of, 11
  recombinant ADF (rADF), 101, 103, *103*, 109, 120
  redox control
    gene expression, of, 115, 118
    signal expression, of, 115, 118
  sequence homology with *E. coli* thioredoxin, 8
  soluble factors related to, 108
  sulfhydryl reduction of proteins by, 9
  TNF, cytoprotection against, *104*
Age, 3, 81
Anemia, 24, 39
Antibodies, monoclonal, 5, 45, 64-65
Antigen recognition, 26, 48
Anti-stress proteins, 3
Anti-T cell antiserum, 5
AP-1, 12, 62, 115-116, 118
AP-3, 62
APO-1, 50
Apoptosis. See *Cell death*.
Arthritis, rheumatoid (RA), and HTLV-I, 13, 30-31
Arthropathy, 13. See also *HTLV-I*.
Aspergillus, 22, 26

ATLV, 1
Autocrine growth factor, 51
Autoimmune diseases, and HTLV-I, 2, 13, 71, 126. See also *specific diseases*.

B

Baeuerle, 112, 116-117
Bannai, 111, 113
B cell
  growth factor (BCGF), 109, 125
  proliferation, 52
Bone marrow
  aspiration, 24
  culture of, 59
  resorption, 53
  transplantation, 28
Broome, 107
Budding, 40, 42

C

Calcium concentration
  calcium concentration, 48-49, 62, 64
  leukemic cells of ATL patients, increase in, 47
Cancer, 2, 112, 118, 120, 123
Candida pneumonia, 26
CD2, 59, 107
CD3, 45, 48-49, 59, 90, 107
CD4
  antigen (TH), 2, 6
  CD4, 45, 51, 69, 71, 73, 90, 113
CD7, 45
CD8, 45, 51, 65, 113
CD20, 90
CD23, 111, 120,
CD28 responsive element (CD28 RE), 62, 64
CD45RA, 50
CD45RO, 50, 90
CD71, 47
Cell activation, 42, 47
Cell death, 42, 50, 84, 95, 107, 112, 119, 126
Cell surface
  IL-2Rα chain, expression on, 69, 71, 73-74
  phenotype, typical, 45, 46, 90
Chemotherapy, combination, 27-28
Clinical utility of redox control, 126
Cloning, gene, 12, 17, 62, 64, 90, 109, 116
Clunie, 109(f)
Connective tissue cells, 53
Curran, 116
Cyclosporin A (CysA), 62, 126-127
Cysteine, 111-113, 117
Cytokine

cytokines, 2, 3, 12, 26-27, 31, 50-54, 59, 64, 101, 109, 120
  receptor superfamily, 65
Cytomegalovirus pneumonia, 22, 26
Cytotoxicity, 103, 109, 126

D

Dean, *109(f)*
Debatin, 50
Dendritic cells, 113
Diphtheria, 28
DNA fragmentation, polynucleosomal, 50
Dröge, 108, 112, 117
Dysplasia, 123
Dysregulation of stress responses, 12

E

Early pregnancy factor (EPF), *109(f)*, 110, 125
ELISA, sandwich, 65, 89, 126
Endoh, 126
Eosinophil cytotoxicity enhancing factor (ECEF), 109
Epidemiological studies, 43, 81
Epithelial tissue and papilloma, 123. *See also HPV.*
Epstein Barr virus (EBV)
  B lymphocytes, and, 120-121
  EBV, 54, 87, 101, 108-109, 111, 115-116, 120

F

Fanger, 107
Feline leukemia virus (FeLV)
  AIDS, and, 39
  Feline leukemia virus, 39
  T cell leukemia, and, 39
FK506, 62, 126-127
Flow cytometric analysis, 45, 50, 67, 69, 90, 113
Fujii, 118, 123
Furuke, 127

G

Gallo, 6-7, 39
Gessain, 29
Gibbon ape leukemia virus, (GaLV), 39
Glutathione (GSH), 107-108, *108*, 112-113, 117, 120
GM-CSF, 52, 54, 73, 82
Go, 115
Gustave-Roussy Institute, 108

H

HAM (HTLV-I-associated myelopathy). *See HTLV-I.*
HAM/TSP (HTLV-I-associated myelopathy/tropical spastic paraparesis). *See HTLV-I.*
HAAP (HTLV-I-associated arthropathy). *See HTLV-I.*
HAU (HTLV-I-associated uveitis). *See HTLV-I.*
Heat shock element (HSE), 115
Heavy metals and oxystress, 115
Hegh, *109*, 110
Hepatoma, 123, 126
Hepatomegaly, 21-22
Herzenberg, 113
Hewlett, 107
Hinuma, 7
Hirashima, 109
Hiro, 108
HLA-DR, 47, 53
Holmgren, 101, 109
Homosexuals, 14
Hori, 109, 115, 126
Host
  defense mechanism, 3
  genes, 3
Human immunodeficiency virus (HIV), 1, 3, 40. *See also AIDS.*
  ADF production, and regulation of, 11
  ADF/TRX, depletion of, 111-115
  HIV, 1, 3, 40
  HIV-I LTR, 54, 112-113, 117
  NF-κB dysregulation in, 116-118
Human papilloma virus (HPV), 123
Human T lymphotropic virus (HTLV)
  Human T lymphotropic virus type I (HTLV-I), 1-3, 7, 39-43
    ADF production
      and regulation of, 11
      overproduction of ADF/TRX, 12, 42
    -associated arthropathy (HAAP), 13, 30-31
      clinical characteristics, 30-31
      histopathological findings, 31
    -associated diseases, 30, 54
    -associated myelopathy (HAM), 12-13, 29-30, 127
      incidence of, 29
      laboratory findings, 30
      neurological manifestations of, 29-30
    -associated tropical spastic paraparesis (TSP), 29-30
    -associated uveitis (HAU), 13, 31-32
      clinical features, 32
      etiology, 31, 81
    ATL incidence in, -infected T cells, 81
    ATLA, and, 9
    cell lines, 10, 12, 91, *94*
    clinical features, 12, 29-32
    gene products, 12
    genetic structure of, 41
    HTLV-I, 30-42
    HTLV-I/ATL, structure of, 42

-infected T cells, dysfunction of, 48
integration site, 10
IL-2 production, -infected T cells, in, 51
  IL-2Rα expression and, 71
mediated transformation of T cells, 8
multiple disease phenotype, and, 2
provirus integration site in leukemic cell
  DNA, 50
-related diseases, other, 29-32, 30, 127
-related pulmonary manifestations, 13
sinovial cells, in, 13
tax gene of, 71
T lymphocytes and, 120
tumor necrosis factor (TNF), and, 52
viral RNA, 52, 71, 95
Human T lymphotropic virus type II (HTLV-II)
  ATL, relationship with, 43
  HAM/TSP, relationship with, 43
  HTLV-II, 43
  Human T lymphotropic virus type II, 43
  incidence of, 43
HUT102, 110
Hybridoma cell line, establishment of, 6
Hydrogen peroxide (H2O2) and oxystress, 105,
  115, 117, 119, 126
Hypercalcemia, 21, 27, 51-54, 73

I

Igata, 29
Immune system
  host, study of, 3
  responses and thiol dependency, 107
  retroviral diseases, stress of in, 2-3
Immunofluorescence staining, 45
Immunological dysfunction, 2, 26
Immunosuppressants, 62
Inamoto, 123
Inflammation, 51-52
Inositol phospholipids, 48
Interferon (IFN), 28, 52, 59
Interleukin-1 (IL-1)
  IL-1, 27, 50-54, 82, 108, 120
  IL-1 receptor antagonist, 51
  IL-1α, 51
  IL-1β, 51
Interleukin-2 (IL-2)
  abnormal expression of, 84
  ATL, involvement in, 10
  ATL, treatment in, 28
  autocrine mechanism, 84, 95
  cytokine receptor systems, and other, 9-12
  engraftment and, 89
  gene
    activation of, 48
    upstream regulatory regions, 61
  IL-2, 59-64, 82, 95-96, 110

IL-2/IL-2 receptor (IL-2R)
  affinity to ligand, 65
  signaling, 9, 65
  system in ATL, 59-74
IL-2 receptor (IL-2R)
  IL-2R, 64-65, 69, 122
  IL-2-receptor α chain (IL-2Rα)
    expression of, in ATL cell lines, 9, 25,
      28, 64, 67, 69, 70, 71, 72, 73-74
    expression of in HTLV-I+ cell lines, 42,
      54, 69
    IL-2Rα, 2, 9, 45, 47-48, 52, 64, 120
    IL-2Rα/Tac overexpression, 12, 84
    mouse antibody, reaction with, 28
    schematic structure of, 62
    soluble IL-2Rα (s-IL-2Rα), 70, 71, 72,
      73, 89
    upstream regulatory regions, 63
  IL-2-receptor β chain (IL-2Rβ)
    expression of, in ATL cell lines, 67-68,
      69, 71
    IL-2Rβ, 45, 64
  IL-2-receptor γ chain (IL-2Rγ)
    IL-2Rγ, 64, 71
    dysregulation and relation to ATL
      leukemogenesis, 9-14
    types, 66
  proliferation of natural killer (NK) cells, 51
  proliferation of T and B cells induced, 51, 53
  studies of, 7
Interleukin-3 (IL-3), 52
Interleukin-4 (IL-4), 65, 121
Interleukin-6 (IL-6), 52-54
Interleukin-7 (IL-7), 65
Interleukin-9 (IL-9), 65
Interleukin-13 (IL-13), 65
Interleukin-15 (IL-15), 65
Iron responsive element (IRE), 116
Irradiation, low dose total body, 28
Ischemia, 126
Ishii, 6
Iwata, 107, 113

J

Japan, southern, 5-6, 29
Jurkatt cells, 116, 119

K

Kagoshima district, 13
Ki, 67, 47
Kitaoka, 126
Klausner, 116
Kusama, 123
Kyoto University Hospital, 1
Kyushu, 5, 13, 31

L

LAK activity, 59, 112
Lentivirus, 39-40
Leukemia. *See also ATL.*
   chronic lymphocytic leukemia (CLL), 5
      B cell, 6
         U.S., in, 6
      T cell, 5
         Japan, in, 6
   cutaneous T cell leukemia (CTCL), 6-7, 42
Leukemic cell(s). *See also ATL.*
   infiltration
      into kidney, *88(f)*, 96
      into lung, *88(f)*, 96
      into various organs, 89, 90, *93*, 95
   lines, establishment of, 6
   methods of engraftment of, *85-86*, 87-91
Leukemogenesis
   leukemogenesis, 3, 74, 82, 84, 87, 91
   mechanism of ATL, related to IL-2R dysfunction, 9
Leukopheresis, 28
Ligand receptor interaction, 48
Liver hepatoma, 123
Long terminal repeats (LTRs), 42
Lymphadenitis, 32
Lymph node enlargement, 19, 21, 25, 90
Lymphoid cells, 12, 24, 107
Lymphoma Study Group (LSG), 19, 25, 27

M

Maeda, 10
Malignant diseases. *See also Nonmalignant diseases.*
   viral etiology of, 2, 39
Martin, *109*, 126
Masuda, 5, 103
Mendelian elements, 39
Mitsui, 101
Miyoshi, 6
Mochizuki, 31
Morgan, 7
MT-1 cells, 6, 42, 111
MT-2 cells, 110, 113
Mycosis fungoides, 6
Myelopathy, 12-13. *See also HTLV-I.*

N

N-acetyl cysteine (NAC), 112, 117. *See also Cysteine.*
Nakamura, 103, 119, 123
Nakashima, 119
Nasopharyngeal carcinoma (NPC), 120
National Institutes of Health (NIH), 14
   studies, U.S., 6-7

Natural killer (NK) cells
   enhancement of activity of, 59
   NK cells, 51-52, 59, 87, 101
Nerve growth factor (NGF), 50
Neurological disorders, 39, 54
Neuronal cells, 126
Neutrophilia, 51
NF-AT transcription factor, 62
NF-IL2A-E, 62
NF-IL2RA, 64
Nonmalignant diseases. *See also Malignant diseases.*
   HTLV-I-related diseases, 2, 12, 13
   nonmalignant diseases, 39
   organ-specific disorders, 2
   tissue-specific disorders, 2
Northern blot analysis, 51, 69, 84
Nuclear factor-κB (NF-κB), 12, 42, 52, 54, 62, 64, 71, 82, 112-113, 116-118

O

Oct-1, 62
Okamoto, 117
Oki-No-Erabu-Jima, 5
Osame, 12, 29
Oxystress, 3, 12, *106*, 110-111, 115-116, 119-120, *121*, 126

P

p53 suppressor oncogene, 81
Papilloma and epithelial tissue, 123
Paracrine growth factor, 51, 84
Parathyroid hormone related protein (PTHrP), 27, 53-54, 82
Peripheral blood leukemic cells, 45, 47-48, 51-53, 67, 69, 87, 91, *92*
Peripheral blood lymphocytes (PBLs), 2, 22, 65, 89, 116, 127
Phosphorylation, 119-121, *119, 121*
PMA, 115
PMBC, 112
Pneumocystis carnii, 22, 26, 53
Polymerase chain reaction (PCR), 31, 53, 95
Polymyositis, 32
Proenkephalin, 54
Protein
   kinase C, 48, 62, 64
   -nucleotide interactions, 12, 54, 116, 118, 120
   -protein interactions, 12
   tyrosine phosphorylation, *119*
Provirus, 42, 50
Pseudomonas exotoxin, 28
Pulmonary manifestations, 13, 18, 22, 32-33, 39. *See also HTLV-I.*

R

Rabbit antisera, 5
Radiometric studies on proliferation of ATL cells in vivo, 95-97
Reactive oxygen species (ROI), 110, 112, 126
Redox control
    ADF/TRX, clinical utility of, and, 126-127
    signal and gene expression, of, 115-120. *See also ADF*
Redox dysregulation, 110-115. *See also ATL.*
Redox regulation and ADF. *See ADF.*
Reed-Sternberg type nuclei, 25
Reproductive system, 110, 118, 123
Respiratory illness. *See Pulmonary manifestations.*
Retinoids, 127
Retroviral diseases. *See specific diseases.*
    first discovered in humans, 1
    stress on immune system, and, 2-3
Retroviruses. *See also Lentivirus.*
    AIDS, and, 14
    ATL, and, 14
    biology of, 40-42
    C-type budding of, 40
    life cycle of, 40
    retroviruses, 39-42
    Rous, 39
Reverse transcriptase (RT), 40, 42, 95
Roederer, 116
Rósen, 109, *109*

S

Sabe, 42
Schenk, 118
SCID mice, 85, 87, 89, 90-92, *92*, 94, *95*
Scratchard plot analysis, *68, 69*
Sepsis, 26
Serum
    ADF/TRX, 125-126
    lactate dehydrogenase (LDH), 18-21, 25
    response element (SRE), 64
Sézary's syndrome, 6, 21, 28
Sheep red blood cell (SRBC), 5
Shikoku, 5
Silverstein, 109, *109*
Simian T-lymphotropic virus III (STLV-III), 40
Skin manifestations, 21, 32
SKT-1B, 113
Southern blot hybridization, 17, 90, *92*
Spastic disease and HTLV-I, 12-13. *See also Tropical Spastic Paralysis.*
SRF, 82
Strongyloidiasis, 26
Structural proteins, 42
Surface associated sulfhydryl protein (SASP), 126

T

Tabira, 126
Tac
    expression in ATL cells, 9, 111
    antibody, anti-, 7, 28, 64, 96
    antigen, 2
    overexpression of in ATL cells, 2
Tagaya, 12, 101, *110*
Takasuki, 5
Taniguchi, 115
Tax
    enhancement of transcription, and, 82
    expression of IL-2Rα, and, 42
    HTLV-I, of, *83*
    IL-2 expression, and, 69, 71
    NF-κB complex, and, 71
    oxidative stress, and, 42
    redox regulation, and, 42
    Tax, 31, 42, 52-54, 64, 82, 95
T cell(s)
    activation of, 48, 59
    growth factor (TCGF), 7
    leukemia in Japan, discovery of a peculiar, 5-6
    marker, 5
    memory, 50
    naive resting, 50
    receptor
        α chain gene, 10, 48
        antigen, 48
        β chain gene, 10, 48-49, 90
        CD3 complex, stimulation of, 48, 59, 62
        gene rearrangement pattern, 17
        responsive elements, 62
    signal transduction, 59, *60*
    stress, and continued activation of, 2
Teshigawara, 12, *109-110*
TH cells
    loss of, 3
Thiols, 111-112, 117-118
Thioredoxin (TRX). *See also ADF/TRX.*
    intracellular function of ADF/TRX, 12
    soluble factors related to human TRX (hTRX), *109*
    TRX, 8, 12, 54, *108*, 109, 118, *124*, 127
13-cis-retinoic acid, *124-125*
3B6/IL-1, 109-110, 120, 125. *See also ADF.*
Tonisse, 110
Trials, clinical, 6, 113
Tropical Spastic Paralysis, 13. *See also Spastic disease and HTLV-I.*
TSP (HTLV-I-associated tropical spastic paraparesis). *See HTLV-I.*
Tumor growth factor (TGF), 53-54, 73, 82
Tumor necrosis factor (TNF), 50, 52-53, 73, 82, *103*, 103-104, 116

Tursz, 108-109, *110*, 120
Tyrosine kinases, 119-120, *121*

U
Uchiyama, 7
Ueda-Taniguchi, 118
Uveitis, 13. *See also HTLV-I.*
Ultraviolet (UV) radiation, 115, 118

V
Viral genes, 3
Virus discovery, in U.S. and Japan, 7
Virus-related transformation, 120-124
Virus Research Institute of Kyoto University, 5
Visna virus, 39

W
Wada, 126
Wakasugi, 101, 108-109, *109-110*, 120

Wakita, 118, 123
Waldmann, 96
Wasting syndrome, 52
Weibull distribution, 81
Wells, 110
Wound healing, 53

X
Xanthoudakis, 116

Y
Yamauchi, 107
Yodoi, 5, *110*
Yoshida, 42
YT cells, 42

Z
ZAP-70, 62

MOLECULAR BIOLOGY INTELLIGENCE UNIT
AVAILABLE AND UPCOMING TITLES

- Organellar Proton-ATPases
  Nathan Nelson, Roche Institute of Molecular Biology
- Interleukin-10
  Jan DeVries and René de Waal Malefyt, DNAX
- Collagen Gene Regulation in the Myocardium
  M. Eghbali-Webb, Yale University
- DNA and Nucleoprotein Structure In Vivo
  Hanspeter Saluz and Karin Wiebauer, HK Institut-Jena and GenZentrum-Martinsried/Munich
- G Protein-Coupled Receptors
  Tiina Iismaa, Trevor Biden, John Shine, Garvan Institute-Sydney
- Viroceptors, Virokines and Related Immune Modulators Encoded by DNA Viruses
  Grant McFadden, University of Alberta
- Bispecific Antibodies
  Michael W. Fanger, Dartmouth Medical School
- Drosophila Retrotransposons
  Irina Arkhipova, Harvard University and Nataliya V. Lyubomirskaya, Engelhardt Institute of Molecular Biology-Moscow
- The Molecular Clock in Mammals
  Simon Easteal, Chris Collet, David Betty, Australian National University and CSIRO Division of Wildlife and Ecology
- Wound Repair, Regeneration and Artificial Tissues
  David L. Stocum, Indiana University-Purdue University
- Pre-mRNA Processing
  Angus I. Lamond, European Molecular Biology Laboratory
- ENDOR and EPR of Metalloproteins
  David J. Lowe, University of Sussex
- Intermediate Filament Structure
  David A.D. Parry and Peter M. Steinert, Massey University-New Zealand and National Institutes of Health
- Fetuin
  K.M. Dziegielewska and W.M. Brown, University of Tasmania
- Drosophila Genome Map: A Practical Guide
  Daniel Hartl and Elena R. Lozovskaya, Harvard University
- Mammalian Sex Chromosomes and Sex-Determining Genes
  Jennifer A. Marshall-Graves and Andrew Sinclair, La Trobe University-Melbourne and Royal Children's Hospital-Melbourne
- Regulation of Gene Expression in E. coli
  E.C.C. Lin, Harvard University
- Muscarinic Acetylcholine Receptors
  Jürgen Wess, National Institutes of Health
- Regulation of Glucokinase in Liver Metabolism
  Maria Luz Cardenas, CNRS-Laboratoire de Chimie Bactérienne-Marseille
- Transcriptional Regulation of Interferon-γ
  Ganes C. Sen and Richard Ransohoff, Cleveland Clinic
- Fourier Transform Infrared Spectroscopy and Protein Structure
  P.I. Haris and D. Chapman, Royal Free Hospital-London
- Bone Formation and Repair: Cellular and Molecular Basis
  Vicki Rosen and R. Scott Thies, Genetics Institute, Inc.
- Mechanisms of DNA Repair
  Jean-Michel Vos, University of North Carolina
- Short Interspersed Elements: Complex Potential and Impact on the Host Genome
  Richard J. Maraia, National Institutes of Health
- Artificial Intelligence for Predicting Secondary Structure of Proteins
  Xiru Zhang, Thinking Machines Corp-Cambridge
- Human IgE Regulation
  Jean-Yves Bonnefoy, Glaxo Institute for Molecular Biology-Geneva
- DNA Strand Break Metabolism
  Michael E.T.I. Boerrigter, Harvard University
- Lymphohemopoietic Cytokines: Growth Hormone, IGF-I and Prolactin
  Elisabeth Hooghe-Peters and Robert Hooghe, Free University-Brussels
- Human Hematopiesis in SCID Mice
  Reiko Namikawa and Maria-Grazia Roncarlo, DNA Research Institute
- Membrane Proteases in Tissue Remodeling
  Wen-Tien Chen, Georgetown University
- Annexins
  Barbara Seaton, Boston University
- Retrotransposon Gene Therapy
  Clague P. Hodgson, Creighton University
- Polyamine Metabolism
  Robert Casero Jr, Johns Hopkins University
- Phosphatases in Cell Metabolism and Signal Transduction
  Michael W. Crowder and John Vincent, Pennsylvania State University
- Antifreeze Proteins: Properties and Functions
  Boris Rubinsky, University of California-Berkeley
- Intramolecular Chaperones and Protein Folding
  Ujwal Shinde, UMDNJ
- Thrombospondin
  Jack Lawler and Jo Adams, Harvard University
- Structure of Actin and Actin-Binding Proteins
  Andreas Bremer, Duke University
- Glucocorticoid Receptors in Leukemia Cells
  Bahiru Gametchu, Medical College of Wisconsin
- Signal Transduction Mechanisms in Cancer
  Hans Grunicke, University of Innsbruck
- Control of the Menstrual Cycle
  Hamish Fraser, University of Edinburgh
- Intracellular Protein Trafficking Defects in Human Disease
  Nelson Yew, Genzyme Corporation
- apoJ/Clusterin
  Judith A.K. Harmony, University of Cincinnati
- Phospholipid Transfer Proteins
  Vytas Bankaitis, University of Alabama
- Localized RNAs
  Howard Lipschitz, California Institute of Technology
- Modular Exchange Principles in Proteins
  Laszlo Patthy, Institute of Enzymology-Budapest

Neuroscience Intelligence Unit

Available and Upcoming Titles

- Neurodegenerative Diseases and Mitochondrial Metabolism
  M. Flint Beal, Harvard University

- Molecular and Cellular Mechanisms of Neostriatum
  Marjorie A. Ariano and D. James Surmeier, Chicago Medical School

- Ca^{2+} Regulation in Neurodegenerative Disorders
  Claus W. Heizmann and Katharin Braun, Kinderspital-Zürich

- Measuring Movement and Locomotion: From Invertebrates to Humans
  Klaus-Peter Ossenkopp, Martin Kavaliers and Paul Sanberg, University of Western Ontario and University of South Florida

- Triple Repeats in Inherited Neurologic Disease
  Henry Epstein, University of Texas-Houston

- Cholecystokinin and Anxiety
  Jacques Bradwejn, McGill University

- Neurofilament Structure and Function
  Gerry Shaw, University of Florida

- Molecular and Functional Biology of Neurotropic Factors
  Karoly Nikolics, Genentech

- Prion-related Encephalopathies: Molecular Mechanisms
  Gianluigi Forloni, Istituto di Ricerche Farmacologiche "Mario Negri"-Milan

- Neurotoxins and Ion Channels
  Alan Harvey, A.J. Anderson and E.G. Rowan, University of Strathclyde

- Analysis and Modeling of the Mammalian Cortex
  Malcolm P. Young, University of Oxford

- Free Radical Metabolism and Brain Dysfunction
  Irène Ceballos-Picot, Hôpital Necker-Paris

- Molecular Mechanisms of the Action of Benzodiazepines
  Adam Doble and Ian L. Martin, Rhône-Poulenc Rorer and University of Alberta

- Neurodevelopmental Hypothesis of Schizophrenia
  John L. Waddington and Peter Buckley, Royal College of Surgeons-Ireland

- Synaptic Plasticity in the Retina
  H.J. Wagner, Mustafa Djamgoz and Reto Weiler, University of Tübingen

- Non-classical Properties of Acetylcholine
  Margaret Appleyard, Royal Free Hospital-London

- Molecular Mechanisms of Segmental Patterning in the Vertebrate Nervous System
  David G. Wilkinson, National Institute of Medical Research, United Kingdom

- Molecular Character of Memory in the Prefrontal Cortex
  Fraser Wilson, Yale University

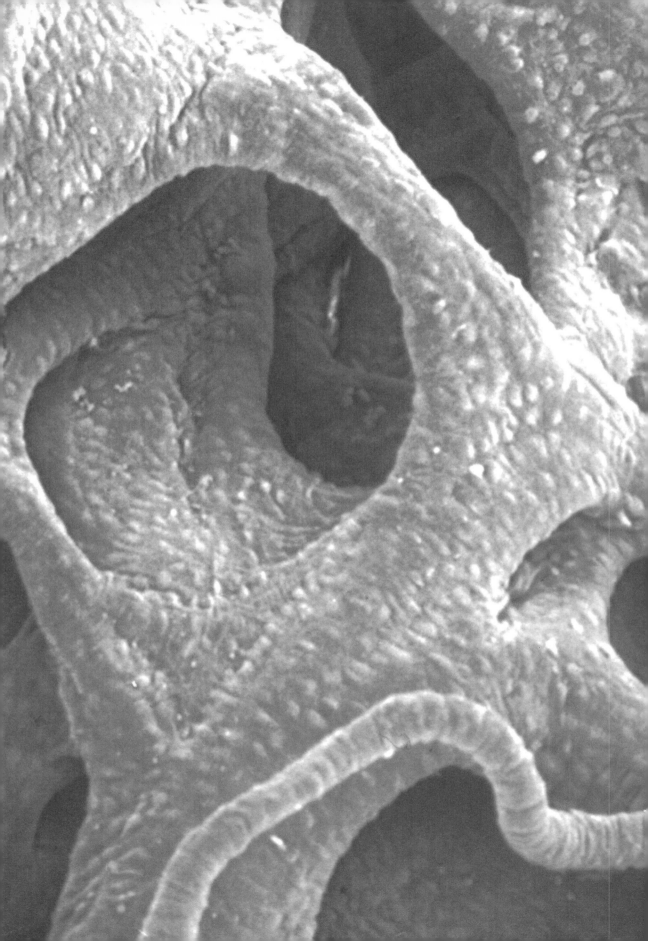